The Last Oil Shock

The Last Oil Shock

*A Survival Guide to the Imminent
Extinction of Petroleum Man*

DAVID STRAHAN

JOHN MURRAY

© David Strahan 2007

First published in Great Britain in 2007 by John Murray (Publishers)
A division of Hodder Headline

The right of David Strahan to be identified as the Author of the Work has been asserted by him
in accordance with the Copyright, Designs and Patents Act 1988.

I

A CIP catalogue record for this title is available from the British Library

ISBN 978-0-7195-6423-9

Typeset in 11.5/14 Monotype Bembo by Servis Filmsetting Ltd, Manchester

Printed and bound by Clays Ltd, St Ives plc

Hodder Headline policy is to use papers that are natural, renewable and recyclable products and
made from wood grown in sustainable forests. The logging and manufacturing processes are
expected to conform to the environmental regulations of the country of origin.

John Murray (Publishers)
338 Euston Road
London NW1 3BH

Contents

Preface

Confessions of an Innumerate Financial Journalist

I ALWAYS THOUGHT geography was unspeakably dull. In school I was so contemptuous of the subject I did no revision and provocatively put my head down and went to sleep during the exam, knowing the invigilators were powerless to intervene. Mrs Couch failed to see the funny side of my score of 9 per cent and demanded a rematch. A few sullen days' revision lifted the tally to 27 per cent, and with honour served I was then allowed quietly to drop the subject.

Only many years later did I discover that geography contains a kernel – geology – that, although based on the study of ancient events, is essential for understanding the modern world. Taken with a dose of maths, it explains the basis of what may turn out to be one of the biggest crises ever to face humanity. How was it that Mrs Couch didn't know this, or couldn't make me see it? It was the mid-1970s, just a couple of years after the first oil shock, when the importance of oil and other natural resources was high on the agenda. More importantly, it was clear by then that an ingenious method for predicting when the oil would start to run short, developed by a Shell geologist named M. King Hubbert back in the 1950s, had already scored its first major success. In 1956 Hubbert predicted American oil production would go into terminal decline within fifteen years, and it did. The implication was clear: the same method could be used to calculate the date on which *worldwide* oil production would inexorably begin to fall. Half a century later, all the evidence suggests that moment is almost upon us.

At maths I proved a natural rather than deliberate duffer. Yet neither this, nor having studied politics at university without a scrap of economics, seemed much of an impediment to a career in financial journalism and broadcasting. So it was I came to be producing the BBC2's *The Money Programme* during the petrol protests of 2000. Casting around for ideas, I saw a letter to *The Times* that piqued my curiosity. By chance its author, David Fleming, lived in my street; there was no excuse but to walk up the hill.

The story he told me was both alarming and persuasive. At the height of the disruption, David argued that the petrol crisis was nothing compared to what was in store. In just a few days a handful of protesters had brought Britain to its knees by picketing the oil refineries. Motorists queued in vain, the supermarket shelves were emptying fast, and the emergency services were within days of collapse. But what if the shortages had been caused not by protesters, soon bought off with a cut in fuel duty, but by geology? What if the day was coming when fuel scarcity would be not only permanent, but permanently getting worse? That conversation led to *The Last Oil Shock*, transmitted in early November 2000.

Since then a sizeable American literature has grown up about this idea, both in print and on the web, much of it deeply apocalyptic. In the shorthand it has become known as *peak oil*, a strangely anodyne term that utterly fails to evoke the catastrophe it is intended to describe. I prefer *the last oil shock*, which is far more suggestive of the severe dislocations the event will probably entail.

Unlike the first two oil shocks, when OPEC (1973) and Iran (1979) deliberately turned off the spigots, this one will have less to do with politics and economics, and everything to do with the iron laws of nature. This will be the *last* oil shock, simply because worldwide oil and gas production are poised to go into terminal decline. It is an idea supported by some of the most experienced oil explorers in the business. Jean Laherrere, who used to be in charge of exploration technique at Total, the fourth largest oil company in the world, puts the situation thus: 'I don't like the thought that oil production will peak and decline any more than I like the thought that I am going to die. But both are facts.' Colin Campbell, whose career spanned Texaco, BP, Amoco, and Fina, says we face a 'major historical discontinuity, by which I mean civilization itself is under threat'. Richard Hardman, the former head of exploration and production for Amerada Hess, agrees: 'People have cried wolf many times before, but I believe this time the wolf really is at the door.'[1]

Less than three years later, as millions protested around the world, it was widely accepted that the invasion of Iraq was 'all about oil', although it was seldom spelled out quite how or in what sense this was true. Insofar as this idea was justified at all, the reasoning was generally based on the well-known 'oiliness' of the US administration, as if the invasion were simply a grotesque payback for the President's friends and backers in big oil. While this may well be part of the explanation, it failed to take account of the administration's publicly demonstrated anxiety about energy security, and it also failed to explain Tony Blair's overwhelming enthusiasm for the attack.

One week into the invasion, in *The War for Oil* (*The Money Programme*, March 2003) I argued that the US-led invasion of Iraq could only be understood in the context of impending worldwide oil shortage. It wasn't corporate greed that was driving events so much as strategic desperation. By then both American and British oil production were in terminal decline, and the fact that this pair of over-the-hill oil producers were also the states keenest on the Iraq adventure seemed unlikely to be mere coincidence.

This idea grew on me as evidence emerged of the profound depths of New Labour mendacity about the war. The lies themselves have been exhaustively dealt with already (Glen Rangwala and Dan Plesch, *A Case to Answer*, or Peter Oborne, *The Rise of Political Lying*), but what has not yet been satisfactorily answered is *why*. Tony Blair is highly intelligent, politically acute and desperate for popularity, and yet the war has made him widely loathed and slashed Labour's majority and tarnished his legacy. He must have been keenly aware of the dangers of being caught out, and yet he lied through his grinning rictus anyway. What reason can have been so compelling as to make him take such risks?

This book is my attempt to answer that question, along with many others that are now more pressing: when will worldwide oil production go into terminal decline; what are the possible outcomes; why do both oil companies *and* environmentalists hate talking about it; does anybody need to worry about global warming any more; will Iran suffer the same fate as Iraq, and for the same reasons; will the oil peak mean financial collapse, economic depression and superpower confrontation; and what, if anything, can we do about it?

When I began my research, the oil peak was still considered something of a cultish obsession not worthy of serious consideration by analysts and economists, but in less than two years the agenda has been transformed. That change has been driven by the soaring oil price, rising turmoil in the Middle East, falling production at some of the world's biggest oilfields, the growth of resource nationalism among producer nations from Russia to Latin America and as utility bills hit our doormats with an increasingly heavy thud. When I started out most senior oilmen dismissed enquiries about depletion with an indulgent smile, but today they are forced to field questions about the date of the peak at almost every news conference. Although many persist in denying the evidence of their eyes, others are beginning to concede the message of the increasingly urgent signals from around the world. Chevron, America's second largest oil company, has launched the website www.willyoujoinus.com to highlight the issues, and its chief executive Dave O'Reilly has concluded that we are seeing the start of a bidding war between East and West for the remaining Middle East oil. Thierry Desmarest, the chief executive of Total, has forecast that oil production will peak by 2020 and – extraordinarily for an oilman – pleaded with governments to suppress demand to delay the onset of decline.

Awareness is spreading outside the oil industry too. Prominent politicians who acknowledge peak oil now include Bill Clinton, Al Gore, former US Secretary of State James Schlesinger, and from Britain former Environment Secretary Michael Meacher. Hollywood is also catching on, providing a surreal image on the podium of an Italian oil depletion conference in 2005, where Sharon Stone was to be found decorously sandwiched between the Libyan oil minister and the Secretary-General of OPEC. Financial markets are beginning to get the message too, with several brokerage firms adopting Hubbert-style analysis, and major institutions launching hedge funds to exploit the crisis. And on the Internet, interest in peak oil is exploding, judging by the number of Google hits: the concept is already twice as famous as the ozone layer, and enjoys a higher profile than David Beckham and Princess Diana combined, although it is only half as popular as the Beatles and receives one-tenth the recognition of climate change.

In the mainstream media, however, the quality of reporting on oil depletion is still wretched, and from governments and officialdom everywhere there is almost overwhelming ignorance, denial and duplicity. The British government in particular squandered a perfect opportunity when it failed to confront the issue in its 2006 Energy Review, an omission that given the strength of the evidence can only be regarded as deliberate suppression. As a result, most people remain ignorant of the enormous dangers we face. This story is now more urgent than ever.

London, November 2006

I

Sources in Washington

Dallas

THE ANSWERPHONE MESSAGE is a lively elderly woman's voice telling me she and her husband are out looking for the pot of gold at the end of the rainbow and if they find it, and if I leave a message, they will share. I laugh out loud; it seems somebody may have taken this literally. It's her husband I'm after, Louis Christian, a retired oil company geologist who lives in Dallas, and a map maker with long experience in the Middle East. In the spring of 1998 the United States Geological Survey (USGS) called to ask him to produce something on Iraq. His new maps would not tell the Marine Corps how to get to Baghdad, but may have everything to do with the reason they went. These were geological, subsurface structural maps, showing the thickness and types of rock at various depths, essential for trying to calculate how much oil might be present. Louis drew them by hand, using information trawled from obscure technical papers in a dozen different languages, and charged a decent whack. That's his business. Over the next eighteen months he would deliver half a dozen. The USGS needed them to complete an assessment of Iraq's potential oil resources, part of a wider assessment of the whole world it was conducting at the time, to be published in 2000. Just routine, apparently. Which makes it all the more interesting when you find out who paid for it.

London

In mid-November 1999, just one week after Christian had delivered the last of his Iraq maps, Dick Cheney flew to London to give a speech at the Savoy Hotel. His audience consisted of 400 oil executives, gathered for the annual lunch of the Institute of Petroleum. By some accounts the occasion had a chilly edge, heightened by Cheney's generally bloodless demeanour, and by the quietly menacing US security men scattered among the diners, some sporting a bulge under their jacket. But these were no goons: 'They were incredibly well informed about oil and strategic issues,' recalls one of those present, 'and spent most of the lunch interrogating us about what we knew.'

Cheney was a relative newcomer to the industry, having spent most of his career as a political fixer, serving under Nixon, Ford and Bush the Elder. As Secretary of Defense he had directed Operation Desert Storm in the first Iraq War. It was only in 1995 that he went into business and became chairman and chief executive of the US oilfield services company Halliburton. Nevertheless his speech showed an acute awareness of the principal fact of life in his latest career: depletion.

> . . . For over a hundred years we as an industry have had to deal with the pesky problem that once you find oil and pump it out of the ground you've got to turn around and find more or go out of business. Producing oil is obviously a self-depleting activity. Every year you've got to find and develop reserves equal to your output just to stand still, just to stay even.[1]

In one sense Cheney's emphasis was not surprising. Oil production in the United States had been in decline – near relentless, geological, terminal decline – since 1970. And a central part of Halliburton's business was the supply of technology to slow – but seldom halt – the falling output from mature oilfields. This was also an issue Cheney's audience grappled with every day, a fundamental fact of life. But as they sat back and digested, Cheney went on to suggest that depletion was becoming an increasingly global problem:

> For the world as a whole, oil companies are expected to keep finding and developing enough oil to offset our seventy one million plus barrels a day of oil depletion, but also to meet new demand. By some estimates there will be an average of two per cent annual growth in global oil demand over the years ahead along with conservatively a three per cent natural decline in production from existing reserves. That means by 2010 we will need on the order of an additional fifty million barrels a day. So where is the oil going to come from?

To this alarming question, Cheney had an immediate answer: 'the Middle East with two thirds of the world's oil and the lowest cost, is still where the prize ultimately lies, even though companies are anxious for greater access there, progress continues to be slow.' One reason for such slow progress, Cheney argued, departing briefly from his script, was another pesky problem: 'US-imposed economic sanctions'.[2]

Cheney had never been much of an enthusiast for sanctions, even against those who committed the foulest atrocities. After Saddam Hussein famously gassed 5,000 Kurds in Halabja in March 1988, the US Senate unanimously passed the Prevention of Genocide Act, which would have imposed a total boycott. According to Peter Galbraith, a senior official at the Senate Foreign Relations Committee who investigated the gas attack, Cheney — at that stage a senior congressman — and National Security Advisor Colin Powell were instrumental in making sure the Act never became law. At the very least, he argues, they did nothing to support it. Perhaps, as honourable men, they were repelled by the implied hypocrisy of the measure; after all, during the 1980s America had supplied Iraq with $5 billion in loan guarantees, weapons, chemical precursors, biological agents including bubonic plague, and even the very crop-spraying helicopters used to mount the Halabja attack.[3] Or perhaps more likely they objected to cutting America off from such a strategically important source of oil, and the hypocrisy was theirs. Galbraith, who later became US Ambassador to Croatia, recalled: 'What I read into it was an indifference to the moral issues that Cheney then trumpeted when making the case for war against Saddam Hussein. Given an opportunity to do something at the time that Saddam was gassing his own people, he did nothing.'[4]

When Iraq invaded Kuwait in 1990, however, sanctions could hardly be avoided, and both the US and the UN imposed a comprehensive embargo which was substantially maintained right up until the second invasion in 2003. This denied international oil companies access to the country with the world's third largest reserves.[5] But at the time sanctions were first imposed nobody worried much about worldwide oil depletion, and, in any event, from the American viewpoint at least, the playing field was level.

3

Nevertheless, US policy towards other 'rogue states' was having some unintended and unwelcome effects. In 1986 President Reagan had imposed comprehensive unilateral sanctions against Libya, the country with the biggest reserves in Africa. In 1995 President Clinton banned US involvement in the oil and gas sector of Iran, with the world's second largest reserves. Realizing that such unilateral sanctions would put American oil companies at a huge disadvantage, the following year he approved the Iran-Libya Sanctions Act (ILSA), intended to penalize any *foreign* company that dared to invest more than $40 million in either country. But ILSA failed miserably from the start. Europe was so outraged by the measure that it threatened a major row through the World Trade Organization, and America quickly backed down.

It must have been particularly galling that as a result of US policy ConocoPhillips was forced to abandon a $550 million contract it had just won in Iran, which was then taken up by Total. The French company went on to win another contract worth $2 billion to develop the massive South Pars gas field, and Clinton was compelled to announce that this and similar deals in future would be given a waiver.[6] None of ILSA's penalties has ever been enforced, despite a growing list of foreign investors, and by 2004 Iran's oil sector had attracted $30 billion from French, Italian and other foreign companies.[7] None of them, of course, was American.

In April 1999, just five months before Cheney's London speech, the United Nations suspended its sanctions against Libya when two suspects in the Lockerbie bombing were extradited for trial, but America's unilateral sanctions remained in place. So while European oil companies vied to win more than a hundred new Libyan exploration licences that were suddenly up for grabs, American companies were again locked out by their own government. US companies that already owned concessions in Libya from pre-sanction days, such as Occidental, Amerada Hess, ConocoPhillips and Marathon, stood to lose them if there was no change in US policy.

No wonder Dick Cheney was exercised about US sanctions. Yet despite observing that oil companies were struggling to replace their reserves, and that American firms were locked out by sanctions from countries that were massively endowed, Cheney professed himself 'unreasonably optimistic' about the future of the industry. This optimism was completely at odds with his entire analysis, unless of course he thought that facts on the ground in the Middle East were about to shift. Cheney concluded: 'Well, the end of the oil era is not here yet, but changes are afoot and the industry must be ready to adapt to the new century and to the transformations that lie ahead. It will mean showing more speed and agility.' And departing again from his prepared text: 'As I have outlined today, there are new areas to co-operate in, new risk, new competition, new roles, new integration and a new convergence with power.'[8]

Four months later, Cheney would experience his own convergence with power: in April 2000 George Bush gave him the job of finding a vice-presidential running mate; in July Cheney himself mysteriously won the contest; and in August he resigned from Halliburton – trousering a multimillion-dollar pay-off – to start the fight for control of America's energy and foreign policy.

North Sea

In November 1999, while Cheney was speaking at the Savoy, an even more significant event was taking place in Britain, or rather underneath its continental shelf. That year British oil production hit its all-time high of 2.9 million barrels per day.[9] From then on Britain's North Sea output would be in relentless, geological, terminal decline, just like America's. It is hard to overstate the importance of this turning point for Britain, and arguably the world, given its evident – though so far unremarked – impact on UK foreign policy. The industry had been expecting the UK peak since the mid-1990s, because the North Sea was becoming what they politely call 'mature'. What may have surprised both oil companies and officials at the Department of Trade and Industry (DTI) was the abruptness of the reversal from growth to decline, and the speed with which oil output began to slide.

They were already worried about the North Sea, but for a different reason. In 1998 as a result of the Asian economic crisis, worldwide demand for oil had fallen sharply, and the average price of Brent crude that year slumped to less than $13 per barrel.[10] Since the average *cost* of producing oil from the North Sea was almost $11 per barrel, the outlook for the UK industry was bleak. So the government promptly pulled a handbrake turn in tax policy, ditching proposals in Gordon Brown's first budget to increase tax on oil companies operating in the North Sea, and introduced new tax incentives instead.

The DTI also set up an Oil and Gas Industry Taskforce, which reported in September 1999. Although its main focus was to see what could be done to mitigate the effects of the price collapse, the Taskforce report showed a clear understanding of the geological maturity of the North Sea. A combination of falling reservoir pressures and the fact that the biggest fields had been discovered and exploited first, meant output was bound to go into terminal decline. A technical annexe to the report predicted that the UK production peak would occur 'within the next two years' and that current production levels were unsustainable 'whatever the oil price'.[11] Nevertheless, energy minister Mike O'Brien wrote confidently: 'The Taskforce's efforts, past, present, and future will slow and delay the decline very significantly.'[12] But by the time those words were published North Sea crude production had in fact already peaked, and would soon begin to fall in earnest. By February 2000 the UK oil industry journal of record, the *Petroleum Review*, was carrying headlines such as 'North Sea Oil Topping Out', and by the end of the year average output had already plummeted by more than 8 per cent.[13] At that rate, production would more than halve in ten years, making Britain critically dependent on foreign supplies, and threatening a severe deterioration in its balance of payments.

Iraq

The collapse of the oil price was also affecting other parts of the world, notably Iraq, but for quite different reasons. Unlike oil production in America and Britain, Iraqi output was a very long way from geologically imposed terminal decline; its status as the country with the world's third largest oil reserves was re-confirmed in June 2000 when the United States Geological Survey released its *World Petroleum Assessment*. The USGS work on Iraq was based on Louis Christian's maps, which were fed into the USGS computer model along with other data to help estimate the amount of oil 'yet to find'. The number they came up with was 45 billion barrels which, added to Iraq's proved reserves of 112, made a total of 157 billion barrels.[14] Only Saudi Arabia and Iran had more. Of the top three, Iraq's oil riches were certainly the least exploited. Since oil exports resumed in the mid-1990s, Iraqi production had averaged less than 2 million barrels a day. But given the scale of the country's reserves, theoretically it had the capacity to produce perhaps three times as much.[15] The reason it wasn't doing so had everything to do with UN sanctions.

The murderous history of UN sanctions against Iraq is already well documented, although now largely forgotten in the shadow of subsequent events.[16] Within a week of Saddam Hussein's invasion of Kuwait, the UN had declared the most comprehensive set of sanctions ever imposed. With the exceptions of medicine and food, nothing was to cross Iraq's borders, and even food was permitted only 'in humanitarian circumstances', whatever that distinction was supposed to mean.[17] In a country that relied on oil exports for 60 per cent of its national income and which imported two-thirds of its food, the effects were predictably catastrophic.[18] The economy shrank by two-thirds within two years, and widespread starvation was only prevented by the prompt imposition by the Iraqi government of food rationing. Even so, by the mid-1990s the average Iraqi's food intake had dropped to just 1,200 calories a day, probably half what you and I consume. By 1998 UNICEF had concluded that a million children under five were malnourished, and the Red Cross that 70 per cent of Iraqi women were anaemic.[19]

The start of the oil-for-food programme in 1997 did gradually improve the Iraqi diet: by 2002 average calorie intake was back up to 2,200 per day. But at the same time the behaviour of America and Britain on the UN Sanctions Committee, which vetted all imports under the scheme, visited many other kinds of misery on the Iraqi population. France, Russia and China seldom objected to Iraqi imports but America and Britain routinely blocked a wide range of humanitarian goods, claiming they were 'dual use'. The list of items thought to have dangerous military applications included baby food, agricultural equipment, X-ray machines, ambulances, wheelbarrows and medicines, from specialist cancer drugs to common painkillers. Carne Ross, the former First Secretary of the British mission to the UN who later resigned in protest over British policy on Iraq, recalls that the US position was particularly restrictive: 'We had American weapons inspectors who really thought that cat litter was being used to stabilize anthrax, that's how paranoid they were.'[20] By July 2002 the value of goods on hold had soared to $5 billion.

The most important health-related 'holds', however, were on electricity-generating and water-purification equipment. During the war the allies had deliberately targeted Iraq's civilian infrastructure, including power stations, water-treatment and sewage plants, and oil facilities.[21] They had bombed Iraq almost back to the pre-industrial age, and now, via the Sanctions Committee, America and Britain would keep it there. By blocking equipment essential for the repair of power, water and sewage plants, and by limiting access to chlorine, the Sanctions Committee forced Iraqis to rely on filthy river water, which caused huge increases in diseases such as typhoid and cholera. In 1998 UNICEF calculated that the death toll in children under five since the war had been *half a million* higher than anticipated. In other words, the combined effects of Allied bombing and sanctions were killing 5,000 small children every month.[22] If this doesn't count as genocide, it's hard to imagine what does, but as Bill Clinton's Secretary of State Madeleine Albright famously said, the price was 'worth it'.[23] Never mind the deliberately uncounted casualties of the second Gulf War and the Iraqi insurgency, from the sanctions alone Clinton, Bush and Blair may have more Iraqi blood on their hands than Saddam. In 1998 the UN official in charge of humanitarian relief in Iraq, Assistant Secretary-General Denis Halliday, resigned in disgust, condemning the oil-for-food programme as 'a Band-Aid for a U.N. sanctions regime that was quite literally killing people'.[24] His successor followed shortly after. But I digress. By now the programme was not just killing people, it was also damaging something far more valuable.

In early 1998 the UN raised the amount it would allow Iraq to earn from its oil exports to fund the oil-for-food programme to $5.2 billion every six months. But Iraq protested that it could never pump enough oil to fund the enlarged budget, because the oil price had collapsed and because of the terrible state of its oil industry. Security Council members suspected Saddam was lying and ordered that a multinational team of experts should be sent to investigate.

One Friday morning in March 1998, British petroleum engineer Paul Wood received a message to fax his CV for UN Secretary-General Kofi Annan to read over the weekend. It came from the boss of Saybolt, a specialist Dutch oil cargo inspection firm which had the contract to monitor Iraqi exports under the oil-for-food programme. Now the company had been asked to put the new investigation team together in a hurry.

It was an impressive résumé. As a young offshore operations manager with Shell in 1971 Wood had been closely involved in the discovery of the massive Brent field in the UK North Sea. Then followed three decades of international assignments including Canada, Barbados, Myanmar, Venezuela, Nigeria and almost every country in the Middle East. Evidently it convinced the Secretary-General; by the middle of the following week Wood was in Cyprus boarding a UN-operated Antonov 28, the only plane allowed in and out of Iraq, bound for Baghdad.

Wood spent the noisy and uncomfortable flight jammed between boxes of cargo and getting to know his new colleagues, a mixed group including oilmen from France, Russia, Jordan, the Netherlands and Norway. Each was an expert in some aspect of downstream production, such as pipelines, storage, or refinery. The team was led by another Briton, Graham Brett of Saybolt, but Wood was the only petroleum engineer on board with upstream development and production experience, and thus the only member qualified to assess the condition of Iraq's oil reservoirs.

During their ten-day trip the team found an oil industry still ravaged by a decade of war – first against Iran and then the allies. At site after site they saw production equipment riddled with bullet holes, and blown-up oil storage tanks with their tops peeled open in the shape of some grotesque flower, blackened sand all around. The country's biggest refinery, Beiji, had been hit by five cruise missiles and 3,000 scatter bombs, yet somehow it still continued to operate.

In addition to the unrepaired war damage, the team discovered a crippling shortage of essential spare parts, routinely denied to Iraq by the Sanctions Committee as 'dual use'. During an interview at his home in Hampshire in May 2005, Wood told me: 'Those poor guys who ran the Iraqi oil industry had been surviving on a shoestring for years.'

For example bentonite, the clay mineral used by oilmen the world over to make 'drilling mud', which lubricates the drill bit and prevents dangerous oil and gas blowouts, was forbidden. So too were the explosive perforation charges which open up the bottom of a well to allow the oil to enter the pipe. As a result only four wells were drilled in Iraq between 1997 and 2001. By comparison, in the British North Sea during the same period the total number of wells drilled was 1,261.[25] The restrictions on water-purification equipment and chemicals also affected the Iraqi oil industry; if you are going to inject water into a reservoir to maintain pressure and keep the oil flowing, it has to be clean, since any particles will clog the tiny pores in the reservoir rock, and this can severely and permanently inhibit production. Yet critical chemicals and pumps were prohibited. Water injection was also hobbled by UN holds on the high-tech meters needed to measure pressure, flow rate, temperature and salinity at the bottom of each well; without them Iraqi reservoir engineers were effectively working blind. Speaking on condition of anonymity, another on the team explained, 'You can't manage what you can't measure.'

The experts' report to the Security Council concluded that the state of Iraq's oil industry was 'lamentable', and was unlikely even to earn the $4 billion that the Iraqis themselves thought possible; $3 billion was more probable. They also warned starkly that 'A sharp increase in production without concurrent expenditure on spare parts and equipment *would severely damage oil-containing rocks* and pipeline systems, and would be against accepted principles of "good oilfield husbandry"' (my emphasis).[26] The experts recommended that $300 million in spare parts be made available, and Iraq went ahead with its plans. Opening all the taps, pumping flat out, in November 1999 Iraq briefly reached 3 million barrels a day.

In January 2000, two years after their first visit, the experts returned to Iraq and were appalled by what they found. True to form, only a fraction of the necessary spares had in fact been delivered, and the temporary spurt in production had only been achieved by short-term, high-risk production strategies, which in some cases were 'entirely run on guesswork'.[27] Despite the Iraqis' efforts, output was now falling inexorably, already down 10 per cent from its recent peak. The UN team reported that production was likely to continue to fall at between 5 per cent and 15 per cent a year, largely the result of Anglo-American obstruction. Wood later told me during our interview in 2005, 'It began to look as if the American and British representatives on the Sanctions Committee had been briefed to do as much damage as possible.'

In their desperation to raise output the Iraqis were using untreated and unmeasured water injection at both their developed supergiant fields. At Rumaila in the south, corrosion and the badly managed water-injection programme had forced the permanent closure of more than sixty wells. At Kirkuk in the north the experts noted: 'With no telemetry, nor operational communications, nor adequate water-treatment spares, the possibility of irreversible damage to the reservoir of this super-giant field is now imminent.' Wood recalled, 'What they were doing threatened to destroy the reservoirs.'

In their second report to the Security Council, Wood and his colleagues made clear that their earlier fears were being realized.

> Poor oilfield husbandry has already resulted in an *irreversible reduction* in the ultimate recovery of oil from individual reservoirs. Crisis manage-ment will continue to exacerbate the *permanent loss of huge reserves of oil*. The group of experts estimates that some of the sandstone reservoirs in the south may only have ultimate recoveries of between 15% and 25% of the total oil that could be drained, because of inadequate water-drive facilities. The industry norm for analogous reservoirs in other countries is in the 35% to 60% range [my emphases].

In other words Wood and his colleagues were warning that, as a result of the sanctions regime, up to half of the otherwise recoverable oil would be permanently lost – no matter who eventually controlled it.

So by April 2000, when Kofi Annan circulated the experts' second report to the members of the Security Council, American and British officials would have been aware that maintaining sanctions against Iraq would:

1. restrict current Iraqi oil production to far below its potential,
2. cause production to fall by as much as 15 per cent a year from then on,

3. permanently damage its reservoirs, and quite possibly halve the total amount of oil that would ever be recovered from the country's currently producing fields.

Officials would also have known that although Iraq's current fields were in danger of being wrecked, the country had a lot more oil in reserve. In yet another report to the Security Council Wood described Iraq's many known but undeveloped fields as 'the largest unexploited collection of super-giant oilfields in the world'. [28] And then there was the USGS estimate of 45 billion barrels yet to find, although under sanctions nobody could go out and look for it. So the prize glistened, just out of reach.

Houston

That same spring, Matt Simmons took a call in his fiftieth-floor office in downtown Houston. It was a young staffer from Governor Bush's office asking him to help with the forthcoming presidential election campaign. Not to lick envelopes, you understand, but to work up some policy proposals for the governor's manifesto. 'I understand you have some concerns about energy?' enquired the staffer. 'How long have you got?' thought Simmons. Forty-five minutes later he was on the team. When Governor Bush's 'Comprehensive Energy Plan' was launched later that year, Simmons had vetted every word of it. This was the document that after the election would form the basis of the work of the Cheney Energy Taskforce.

Simmons is an energetic, ruddy-faced man, who for over thirty years has served as banker to the Houston oil services sector, financing companies' expansion in the good years, restructuring them in the bad. So far the firm's deals total $77 billion, and today Simmons & Company is the world's largest energy investment bank. Simmons is also something of a contradiction. On the one hand, from his background and business interests he is a natural Republican; on the shelf behind his desk sits a mounted photo of him and his wife with George Bush at a barbecue in Crawford. And the Republican instinct when faced with problems of energy supply is to go out and find some more – even if it *is* buried under some pristine wildlife reserve. On the other hand, he is also what you might call a paid-up member of a rather different group, many of whom are distinguished scientists and highly experienced oil explorers, who warn that the world is barrelling towards the mother of all energy crises: peak oil, or the last oil shock.

Simmons, who travels the world almost non-stop to give talks and papers, regularly shares a public platform with members of this group, and by 1999 was already starting to share their bleak worldview. A sample of his talks from the time gives a flavour of his concerns: 'Raising the Spectre of an Oil Shortage' (October 1999); 'Is Oil Price Shock Around the Corner?' (January 2000); 'Revisiting the Limits to Growth: Could the Club of Rome Have Been Correct, After All?' (October 2000).[29] In fact one of the few significant differences between Simmons and the rest was that while they were considered voices in the wilderness, he was *connected*. And now their shared view of the world energy predicament was being fed directly into the Bush election team. 'What I basically told them is that there were some looming energy problems, because we'd run out of productive capacity,' he later told me. 'In the nineties we had basically used up all of the cushions, and yellow lights were going off all over. We were barrelling into a really nasty energy crisis.'[30] The message evidently found a ready audience in the would-be administration. Simmons wasn't to know that some already had their own ideas on how to fix the problem.

Clwyd

New Labour's re-education on the critical importance of oil began in a cattle market in the tiny Welsh cathedral town of St Asaph. On the evening of Thursday 7 September 2000, about 150 angry farmers, lorry drivers and cabbies met to discuss the price of fuel, which had risen more than a fifth over the previous two years. The newspapers that week had predicted that prices would soon hit £4 a gallon. The rise was due to the spiralling cost of crude oil on world markets, but since almost three-quarters of UK fuel prices consisted of tax, the meeting's anger was largely directed at the government. Only days before, French fishermen had wrung concessions from their government by blockading the country's harbours, and somebody suggested a protest at the oil refinery at Ellesmere Port. So a motley convoy of cars, trucks and tractors set off up the A55.

The protests spread like fire in a fuel depot, organized by CB radio and mobile phone. Soon refineries up and down the country that usually sent out a tanker every three minutes were down to one an hour, and petrol stations, which rely on regular 'just-in-time' deliveries, were starting to run out. Mile-long queues sprang up as people panicked and fuel hoarding set in. By the weekend − just two days after the protests started − the headlines were stark: 'Petrol Pumps Will Run Dry, Says Shell'; 'Protestors Cripple the Oil Refineries'; and 'Panic as Garages Run Dry'.

Tony Blair, on a tour of the north to promote the wonders of the dot.com economy, found the old, real economy grinding to a halt all around him. In Hull he was greeted by fuel protests and go-slow convoys, and even forced to cancel dinner. Energy minister Helen Liddell was despatched to visit the Queen at Balmoral to be granted emergency powers. And deep below Whitehall in COBRA (Cabinet Office Briefing Room A), the government's crisis management committee convened. Its first briefing note to Blair was stark: 'Oil industry reports that the situation is near breaking point. There had been no significant breakouts from the oil refineries, forcing the MoD to look at options for military assistance.'[31]

Behind the scenes it was chaos. According to a businessman who was closely involved, officials 'were running around like headless chickens, it was appallingly badly handled'. Whitehall officials planning which petrol stations to use as fuel centres for the emergency services discovered their data was three years out of date. Plans to use army HGV drivers foundered when it was realized there wasn't time to give them the safety training essential for loading and discharging tankers. On Tuesday morning up to a third of all petrol stations had closed, Avon Ambulance Service announced it had just two days' fuel, and the *Sun* headline screamed: 'We'll Run Dry Today'.

In an interview with ITN David Blunkett likened the crisis to the winter of discontent, the national strike which heralded eighteen years in opposition for Labour. A Downing Street aide 'visibly winced' as he said it, but Blunkett can only have been voicing the fear stalking every government minister. The cause was different but the effects strikingly similar: rubbish was again being left uncollected in the streets and there were already concerns that, like twenty years before, the dead would soon be left unburied. As former Environment Secretary Michael Meacher told me, 'We were peering into the abyss.'

The suspicion grew that the oil companies were conniving in the crisis. By now at most refineries the police had cleared the gates and protesters were held back at the sides of the road, but still the tankers failed to roll. The companies didn't like the high UK fuel duty any more than the protesters, and were evidently just playing dead. Blair personally hit the phones to oil company bosses, and on Tuesday evening claimed that everything should be getting back to normal within twenty-four hours. 'Everything is now in place to get the tankers moving. The oil companies are agreed that they must move supplies.' But *still* the tankers failed to roll.

Wednesday morning came with the headline: 'The Day the Fuel Ran Out'. But it wasn't just the fuel. Common decency was wearing thin as scuffles broke out on garage forecourts, and as thieves stole petrol from nurses' cars in a hospital car park. For many work was also disappearing, as factories started to lay off employees, and Jaguar, Ford and Vauxhall announced that they would have to stop production by the weekend. Most important of all, supermarket shelves were being stripped of staple foods in scenes of panic buying. Sainsbury, Asda and Safeway reported that some branches were having to ration bread and milk. A spokeswoman for the British Retail Confederation said retailers had fuel to keep up deliveries for 'the next twenty-four to forty-eight hours or so'. Sir Peter Davis, chairman of Sainsbury, wrote to Tony Blair warning him that food would run out in 'days rather than weeks'.

It was in this desperate atmosphere that Tony Blair called the oil bosses in to Number 10 one by one to read them the riot act. At the same time government ministers hit the TV studios with the message that 'lives were at risk'. Public support for the protesters began to ebb, and in the small hours the pickets were called off. Just as the tankers finally started to roll on Thursday morning some of the oil companies

15

decided, with exquisite timing, to announce a price rise. Now Blair was incandescent, and hauled all the oil company bosses in together for a collective carpeting. Downing Street refused to release the minutes of this meeting under the Freedom of Information Act, on the basis that to do so would be bad for policy formation because it would inhibit ministers, officials and others from expressing themselves 'freely and frankly'.[32] I am not surprised; sources suggest that on this occasion the air was blue with free and frank discussion.

Evidently a deal was cut. The oil companies were persuaded to keep prices on hold until November, when Gordon Brown announced a cut in duty on ultra-low sulphur diesel – a sop to the hauliers, cleverly dressed up as a green measure. Shortly afterwards the oil companies jacked up prices by three pence a litre, totally negating the effect of the tax cut. So the protesters had been neutralized, the Chancellor given time to come up with a concession that he could spin as being part of the normal budget process, and after a brief delay the oil companies got their price rise.

Despite the neatness of the political fix, nothing could hide the fact that Britain had gone from normality to incipient hunger within the space of a single week. The weekend headlines reflected the severity of the crisis – '48 Hours From Meltdown' – and its political impact – 'Seven Days That Shook New Labour'. For a party determined to win a historic second term, and with the election less than a year away, the polls were grim. Labour's support had collapsed from 51 per cent a month before to just 36 per cent, and for the first time in eight years the Tories were ahead.[33] The vast majority blamed the government rather than the protesters for the crisis. None of this had anything to do with geologically imposed oil depletion, in the North Sea or anywhere else, but it cannot have failed to impress on Tony Blair the fundamental importance of energy security.

Washington

Three months later the oiliest administration ever took power in the US, with the help of $27 million in campaign contributions from the industry, although the calibre of some of the former oilmen now in charge should not be exaggerated.[34] Unlike his father, George W. Bush was an oilman only in the sense that Eddie the Eagle was a skier. According to Craig Unger's *House of Bush, House of Saud*, in the late 1970s and early 1980s his oil company Arbusto 'drilled one dry hole after another'. He was even forced to change the firm's name – the Spanish word for bush – by the derision of competitors, who spotted the opportunity for a heavy pun on its financial condition. Bush had to be bailed out several times, first by Harken Energy, which gave him a generous price and a continued seat on the board, and later by Saudi investors connected to the Bank of Credit and Commerce International, BCCI.

Some of the most powerful voices in the new administration were rather better qualified. Cheney as we know had run one of the world's biggest oilfield services companies, and National Security Advisor Condoleezza Rice had served on the board of Chevron for a decade. It's well known that the company named a tanker in her honour, although it was re-christened *Altair Voyager* after she joined the administration. What is less widely appreciated is that towards the end of her time at the company, in 1998, Chevron's oil production began to slide into what looked for a time like terminal decline. Every year until 2005 its production dropped, despite the company investing billions of dollars in exploration and production.[35] So Rice also knew all about depletion.

Defense Secretary Donald Rumsfeld was not an oilman as such, but owned several million dollars in oil company shares, and far more importantly had helped to develop the Neocon agenda on oil and the Middle East. Rumsfeld was a key supporter of an ultra-right-wing thinktank called the Project for a New American Century (PNAC), whose ambition was to maintain America's military and economic domination of the post-Cold War world. Other supporters included Cheney; Deputy Defense Secretary Paul Wolfowitz; chairman of the Defense Policy Board Richard Perle; Elliott Abrams, an official on the National Security Council; and John Bolton, the 'quintessential kiss-up, kick-down sort of guy' whom President Bush later appointed his ambassador to the United Nations.[36] PNAC believed America was the only superpower now worth the name, and intended to keep it that way: no other power should ever be in a position to challenge US hegemony. So PNAC urged massive increases in defence spending so that US forces could 'win multiple simultaneous large-scale wars'. To do this it would need to maintain its nuclear superiority, deploy global missile defence and, in a truly hellish vision, 'control the new "international commons" of space and "cyberspace" and pave the way for the creation of a new military service – U.S. Space Forces – with the mission of space control'.[37] But you couldn't do all that without fuel. Imagine the embarrassment of achieving total domination of earth, the universe and everything, only to find your starship troopers have run out of gas.

Rumsfeld evidently already had a plan to avoid that eventuality. In 1988 he and other PNAC supporters had written two open letters urging President Clinton to make getting rid of Saddam Hussein a policy priority, arguing that if the Iraqi leader were to acquire weapons of mass destruction 'the safety of American troops in the region, of our friends and allies like Israel and the moderate Arab states, *and a significant portion of the world's supply of oil* will all be put at risk' (my emphasis).[38] However the aim was clearly broader than simply toppling one brutal dictator: 'We should establish and maintain a strong U.S. military presence in the region, and be prepared to use that force to protect our vital interests in the Gulf.' Another PNAC document contends: 'While the unresolved conflict with Iraq provides the immediate justification, the need for a substantial American force presence in the Gulf transcends the issue of the regime of Saddam Hussein.'[39]

Almost the first act of the new administration was to convene the National Energy Policy Development Group, commonly known as the Cheney Energy Taskforce. Almost, but not quite. At the very first meeting of the National Security Council, President Bush directed Rumsfeld to 'examine our military options' on Iraq. Ron Suskind, whose coruscating account of the Bush administration's early years, *The Price of Loyalty*, is based largely on the testimony of former US Treasury Secretary Paul O'Neill, writes: 'The meeting had seemed scripted. Rumsfeld had said little, Cheney nothing at all, though both men had long entertained the idea of overthrowing Saddam. Rice orchestrated, and Tenet [George Tenet, CIA Director] had a presentation ready.' Suskind concludes: 'Ten days in, and it was about Iraq.'

Detailed planning for the invasion started immediately: 'Already by February, the talk was mostly about logistics. Not the *why*, but the *how* and *how quickly*.'

As planning for the invasion gathered pace, so the Cheney Energy Taskforce got down to work, first meeting on 9 February 2001. It seemed to have been structured to maintain maximum secrecy about its proceedings. Unlike many blue-ribbon commissions on policy, which have traditionally tended to co-opt outside experts, the Taskforce was made up exclusively of senior government officials such as Cheney, O'Neill, Secretary of State Colin Powell, and Energy Secretary Spencer Abraham. This meant that the proceedings were exempt from outside scrutiny, which would otherwise have been compulsory under the Federal Advisory Committee Act (FACA). An anti-corruption non-governmental organization called Judicial Watch – slogan: 'because no one is above the law!' – mounted a series of legal challenges, which Cheney fought all the way up to the Supreme Court, and eventually defeated. Apparently he was.

Judicial Watch was concerned about reports that oil industry executives and lobbyists were enjoying secret and privileged access to the Taskforce proceedings. Although its FACA challenge ultimately failed, the group had greater success under the Freedom of Information Act, and forced the disclosure of 39,000 documents. Although many had been heavily censored, the Taskforce papers showed the scale of oil industry penetration and influence. Since taking office Spencer Abraham had met thirty-six energy industry representatives to talk specifically about Taskforce business, and three times that many in total. In contrast, a coalition of environmental groups was refused a meeting with Abraham because of his 'busy schedule'.[40] Now it was obvious why. But it wasn't the fact that Taskforce members had actually met industry executives that was so interesting, as much as the effort they put into covering their tracks. Many of the 'released' documents had been entirely blanked out, and officials admitted that many others had been withheld altogether to keep secret the internal debate leading up to the final report. What did they have to hide?

More than a year later, in July 2003, quite unexpectedly another small batch of documents arrived in a single envelope at the offices of

Judicial Watch in Washington. 'I was looking at them and laughing,' the organization's director Tom Fitton told me eighteen months later, 'thinking oh my gosh, everybody's going to go crazy over these.' There was in fact a total and shameful silence from the mainstream US media when these documents were released, although they spread rapidly over the Internet.

Among the seven sheets of paper were maps of the United Arab Emirates, Saudi Arabia and Iraq (figure 1). Unlike Louis Christian's maps (which we'll come to) these were not geological, but rather more immediately practical, showing supergiant oilfields, pipelines and refineries. The map of Iraq also showed the location of nine virgin exploration blocks covering the vast western desert, which the Canadian columnist Linda McQuaig in her book *It's the Crude, Dude* likened to 'a supermarket meat chart, which identifies the various parts of a slab of beef so customers can see the most desirable cuts'. The papers also contained a list of foreign suitors for Iraqi oilfield contracts, describing sixty-three companies from thirty countries and the status of their negotiations with the Iraqis. It was a truly international collection, with companies from Algeria to Canada, Norway to Vietnam, and large contingents from China and Russia. The Russian company LUKoil had a production-sharing contract (PSC) for the huge West Qurnah field, and France's Total was shown to be close to a deal on the vast Majnoon field, with 'PSC "agreed in principle" January 1997'. None of the companies was American, and only two tiny UK companies were even listed as having 'discussions'.

So in early 2001 while the National Security Council was planning the invasion of Iraq, the Energy Taskforce was poring over maps of its oilfields and lists of foreign competitors, and one of the most powerful men in the administration – Dick Cheney – was deeply involved in both processes. And we are invited to believe that the invasion had nothing to do with the oil.

Any faint doubts that the two policies were utterly entangled were finally despatched by a revealing memo, dated 3 February 2001, which was later leaked and reported by the *New Yorker* magazine:

> The top-secret document, written by a high-level N.S.C. official, concerned Cheney's newly formed Energy Task Force. It directed the

Iraqi Oilfields and Exploration Blocks

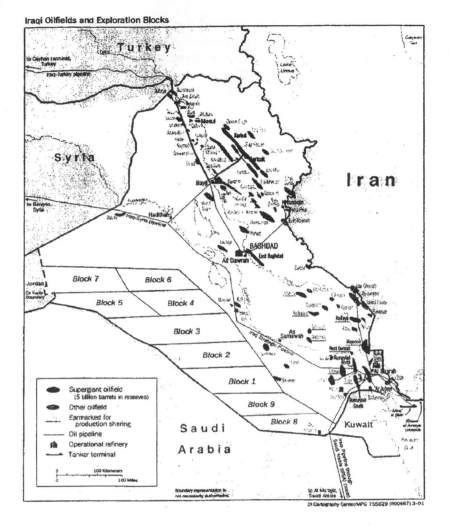

FIGURE I. The map of Iraq's oilfields and virgin exploration blocks viewed by the Cheney Energy Taskforce. Reproduction courtesy of Judicial Watch

N.S.C. staff to cooperate fully with the Energy Task Force as it considered the 'melding' of two seemingly unrelated areas of policy: 'the review of operational policies towards rogue states,' such as Iraq, and 'actions regarding the capture of new and existing oil and gas fields.'[41]

It was no longer necessary to surmise the connection; administration officials themselves had made it explicit.

Amid all the excitement about secret meetings and memos, nobody paid very much attention to the analysis contained in Cheney's report, published in May 2001, which is itself illuminating. The entire document is infused with a sense of crisis and insecurity about energy supplies. A graph showed the widening crevasse between declining US oil production and rising demand, and the report noted: 'we produce 39 percent less oil today than we did in 1970, leaving us ever more reliant on foreign suppliers. On our present course, America 20 years from now will import nearly two of every three barrels of oil – a condition of increased dependency on foreign powers that do not always have America's interests at heart.'[42]

Of the report's conclusions, only the proposal to open the Arctic National Wildlife Refuge (ANWR) to drilling generated much public debate – indeed, uproar. Others went largely unremarked but were rather more significant. Three key recommendations were: to 'make energy security a priority of our trade and foreign policy'; to conduct a 'comprehensive review of sanctions' in which 'energy security' should be one of the factors, which clearly implied trying to find ways to lift the various embargoes; and to encourage Middle Eastern oil-producing countries to 'open up areas of their energy sectors to foreign investment'. At point only a very few people understood quite how they meant to achieve those ends.

London

By now Britain's energy vulnerability was also becoming obvious as North Sea output continued to plummet, and in May 2001 the DTI set up a specialist marketing team whose job was to try to slow the decline. With many fields now severely depleted, major companies were beginning to scale back their North Sea investment. So Jim Munns – a former executive with Texaco and Amoco – got the job of travelling the world's oil investment centres to drum up interest from smaller companies for whom it might still be worth investing. This meant clocking up some airmiles. At meetings from Houston to Calgary to Aberdeen, he talked up the opportunities that still existed for smaller players, but it was clear from his presentations that the overall outlook was grim. One of his slides, produced by DTI officials, showed that even with the most positive assumptions North Sea oil production from existing fields would plunge from almost 2.9 million barrels per day in 1999 to less than 500,000 by 2020, with gas not far behind.

Despite this precipitous outlook, Jim Munns recalls no panic or alarm among his colleagues at the DTI. The assumption was that Britain's security of supply was not threatened, and that the world market would provide for the country's needs: 'It wasn't a big issue.'[43] Subsequent events suggest that this surprisingly relaxed view may not have been shared in Downing Street.

By now both the American and British governments had good reason to be nervous about energy security, but only the US administration had expressed its doubts, both openly and inadvertently. On the face of it British officials were still not in the least concerned. Even two years later, on the eve of the invasion of Iraq, a Department of Trade and Industry White Paper entitled *Our Energy Future – Creating a Low Carbon Economy* would dwell overwhelmingly on the challenges of cutting CO_2 emissions rather than those of securing oil supplies. It noted that Britain would become a net importer of gas by 2006 and of oil by 2010, but concluded that there was little to worry about since other countries would want to sell to us. Neither was the world about to run out any time soon: 'Globally, conventional oil reserves are sufficient to meet projected demand for around 30 years, although new discoveries will be needed to renew reserves. Together with non-conventional reserves such as oil shales and improvements in technology, there is the potential for oil reserves to last twice as long.'[44] As far as one can tell, DTI officials actually believed this complacent twaddle. The question is, did Tony Blair?

Crawford

Seven months on from 9/11, the Prime Minister stepped down from the presidential helicopter in Texas to be greeted warmly by President Bush. It was a critical weekend for both men: Blair under intense pressure from his backbenchers; Bush impatient to invade, but lacking the allies he needed for political cover. The summit had been preceded by weeks of intense diplomatic traffic and a flurry of confidential memos. Mr Blair himself had held talks with Dick Cheney at Number 10 the previous month, his foreign policy adviser Sir David Manning had met Condoleezza Rice and her NSC team, and Britain's US Ambassador Sir Christopher Meyer had lunched discreetly with Deputy Defense Secretary Paul Wolfowitz.

Although both Bush and Blair were accompanied by the usual cadre of senior officials, during the summit the two men 'spent long periods alone together' when their discussions reportedly went unminuted.[45] On the Saturday morning, unusually, Blair was invited to sit in on Bush's daily briefing by the CIA. At a press conference Blair declared, 'It has always been our policy that Iraq would be better off without Saddam Hussein,' and in a speech he uttered for the first time the words 'regime change' in the context of justifying the war.[46]

While Tony Blair would continue to claim publicly that the decision to go to war had not yet been taken, behind-the-scenes briefings made clear that there had been a deal, neatly summed up by *The Times*: 'Iraq Action is Delayed But "Certain"'.[47] Bush agreed to what would clearly be a lengthy attempt to drum up some kind of international legitimacy, and in return Blair committed Britain to take part in the attack. The bottom line was clear in the Sunday headlines: 'Blair to Back US War on Saddam'; 'Blair Agrees on Need to Oust Saddam'; 'Blair Pledges War on Iraq'.

But that wasn't all they talked about at the summit. They also talked about oil. On the very weekend that Bush and Blair forged a pact to invade Iraq, they also quietly set up a permanent diplomatic liaison called the US–UK Energy Dialogue, whose main aim was to secure 'energy security and diversity'. According to documents released to me under the US Freedom of Information Act, it was Blair who proposed the idea, to which Bush agreed.[48] Unlike the deal on Iraq, this agreement was kept secret, and its existence was only uncovered later by Rob Evans and David Hencke of the *Guardian* newspaper through their own Freedom of Information Act enquiry in the US.[49]

Over the following year the Dialogue met at least another eight times, either at ministerial or senior official level in Washington and London. In preparation for the first ministerial meeting, Trade

Secretary Patricia Hewitt wrote to US Commerce Secretary Don Evans that she thought the purpose of the Dialogue was 'to bring together the separate strands of our energy policy and its wider foreign policy context', and that 'the primary focus of discussion would be how to ensure/improve energy security and the diversity of supply'.[50] The last meeting took place in late February 2003, just before the invasion. During the war and its immediate aftermath there continued to be informal contacts between US and UK officials, and the Dialogue resumed formally in February 2004 with a transatlantic video conference between senior civil servants.[51]

The Foreign Office has refused to release the minutes of these meetings under the Freedom of Information Act, because 'American officials shared information with their British colleagues in the context of the UK-US Energy Dialogue and related talks on the understanding that it should remain confidential.'[52] How very convenient. The Americans themselves, however, are not quite so sensitive about the Dialogue, and my own Freedom of Information Act enquiries prompted the release of a similar batch of documents to those obtained by the *Guardian*. They include agendas, correspondence, and a progress report to Bush and Blair signed off by US Energy Secretary Spencer Abraham and Commerce Secretary Don Evans in July 2003, but not the actual minutes of the meetings of the Dialogue. Nevertheless, the paperwork gives strong hints of a shared paranoia over energy security.[53]

The Dialogue had been set up because Bush and Blair had decided that 'the US and the UK should together address the strategic challenges of international energy policy', and was based on the 'frank sharing of strategic analysis and assessments'. Its key aims were 'promoting the security and diversity of future international energy supplies', and 'mitigating the risks of increasing global reliance on Middle East oil, *including looking at the investment needs of the region*' (my emphasis; this is a particularly interesting phrase since at this stage Western oil companies were almost wholly excluded from the region, except as hired help). Separate working groups had been formed within the Dialogue to develop joint policy approaches to vital oil producers such as Russia, Central Asia and the Caspian region.

Its tone was markedly different from the blandly reassuring statements from the DTI, and the very fact that Blair had instigated the

Dialogue tended to suggest that he was already convinced that world oil supplies were at greater risk than the British government was prepared to admit publicly. But this should come as no surprise. Tony Blair already possessed both the knowledge and the scars to make him alert to the critical importance of energy. In opposition, he had been New Labour's Shadow Energy Secretary at the time of electricity privatisation, and according to some who worked with him, had shown a masterful grasp of the brief. Once in power, he gained raw and personal experience of energy's power to disrupt; the first, massive, re-election threatening crisis of his premiership had been brought about by a fuel blockade.

So what was it that Blair discussed with Bush and the rest at Crawford in April 2002? Did they mull over the fact that oil production outside the Middle East and Russia had not risen since 1998, deepening the West's dependence on those producers; did they debate whether OPEC might soon find it difficult to keep up with world-wide demand, or might choose not to; did they voice their deepest fears of a Wahabist uprising in Saudi Arabia; did they note US-UK-driven sanctions were wrecking Iraq's biggest fields; did they pore over the lists of foreign suitors lining up for Iraqi oil contracts, none of them American, nor British of any significance, and ponder whether things might turn out differently with regime change; did they con-sider whether toppling one dictator to gain access for Western oil companies might have a regional effect – that an attack on Iraq could be useful *'pour encourager les autres'*? And was Bush's parting shot to Blair, 'Whose side do you want to be on when the oil starts to run short?'

Abuja to Tehran

Whatever transpired at Crawford, the US administration lost no time in trying to develop alternative sources of oil to lessen its dependence on the Middle East, and British officials became closely entwined in the effort. Shortly after the Dialogue was set up, in the summer of 2002, a flurry of American delegations arrived in the Nigerian capital Abuja, with the aim of increasing the country's oil exports to the

United States. From the US point of view there would be several advantages. Nigeria was far closer to America than the Middle East, it was not threatened by the same kind of instability, and the country had unexploited deepwater reserves which could be used to raise its output significantly.

The first group to be received by President Obasanjo included executives of ExxonMobil, and was led by Paul Wihbey, a Neocon lobbyist representing a right-wing Israeli thinktank, the Institute of Advanced Strategic and Political Studies, whose aim was to reduce United States' reliance on Saudi oil supplies. A couple of weeks later the US Assistant Secretary of State for Africa, Walter Kansteiner, arrived for more discussions with the President and other Nigerian leaders.[54] He too wanted to shift American oil procurement from Saudi Arabia and Kuwait to favour West Africa.

The Nigerians listened politely as the Americans suggested various deals to guarantee preferential American access to their oil. One idea was that the US should lease a proportion of Nigeria's reserves to form part of America's emergency supply, the Strategic Petroleum Reserve. The oil would stay underground in Nigeria's oilfields, but in the event of an emergency would be pumped for exclusive American use. Another yet more ambitious plan was that Nigeria should leave OPEC and commit to sell every barrel it produced to America. By leaving the cartel Nigeria would no longer have been restricted by a production quota, and could have increased its output substantially, but even this was not enough to persuade it to sign an exclusive agreement. It was much more in Nigeria's interests to be able to sell to the highest bidder in an open market than to restrict itself to a single customer, especially if prices were to rise in future. Evidently the Americans were trying to get their energy security on the cheap, and wanted to fix the prices Nigeria would receive in advance at a discounted rate. So the 'deals' all died like strays in the streets of Abuja. According to Jonathan Bearman who runs Clearwater, a specialist West African business intelligence company, none of the proposals was ever likely to succeed: 'Naturally the Nigerians didn't want to tie up their supplies in long-term relationships which were going to reduce the amount that they could earn from their oil. Having the US trying to secure crude at a discount on forward buying was not very appealing.'[55] The American approach may have been naïve and poorly thought out, but it was yet more evidence of their energy paranoia.

The Foreign Office refuses to release any documents it may have about this episode, but it seems unlikely that it didn't come up in one of the Dialogue's meetings, held almost monthly at this stage. In any case, British officials were themselves soon involved in trying to maximize African oil production. According to the Dialogue progress report they had been given the job of working up 'investment issues facing Africa that could be ripe for US–UK coordinated attention'.[56]

In February 2003 nobody outside Whitehall or the Washington beltway knew that America and Britain were co-operating so closely over oil. As millions protested around the world against the impending war, the people waving 'No Blood for Oil' placards were unaware that the Middle East and Energy Security working group of the US–UK Energy Dialogue held a meeting in Paris that month. The working group noted that world oil demand was forecast to soar by 45 million barrels per day to 120 million barrels per day in 2030, and that most of this would have to come from the Middle East. 'To meet future world energy demand, the current installed capacity in the Gulf (currently about 23 mb/d) may need to rise to as much as 52 mb/d by 2030.' The working group then proposed 'a targeted study to examine the capital and investment requirements of key Gulf countries'.[57] Of course it may just be coincidence that on the eve of the invasion of Iraq American and British officials were discussing in detail how to raise Middle East oil production. On the other hand, it may not.

Tony Blair denied repeatedly that the war was 'all about oil', but in retrospect his denials inadvertently highlighted one of the several ways in which it may have been. On the eve of the invasion, Blair told the studio audience of the BBC's *Newsnight*: 'If the oil that Iraq has were our concern we could probably cut a deal with Saddam tomorrow in relation to the oil. It's not the oil that is the issue but the weapons.' We know all about the weapons – *what weapons?* – but it's easy to miss the fact that Blair's claim about the oil was also highly tendentious.

The idea that Saddam Hussein would voluntarily strike a deal on oil with his two principal tormentors on the Security Council is laughable. Had sanctions been lifted and Saddam left in power, it was widely understood that the immediate beneficiaries would have been

Russia, France and China. Carne Ross, former First Secretary of the British mission to the UN, recalls: 'We were acutely conscious of the fact that Saddam had carved up all the oil exploration deals and contracts with big non-US and non-UK multinationals.'[58] So for the US-UK axis the conundrum was this: Iraqi production could only be raised – or even stopped from collapsing – if sanctions were lifted, but lifting sanctions without regime change would benefit their international rivals. It followed that the only way to raise Iraqi output, fend off the global oil peak, *and* keep some kind of US-UK stake in the new Iraqi production was to depose Saddam Hussein. This analysis is supported by the fact that there was at the very least a general understanding between the UK-US axis and the Iraqi opposition groups that regime change would also be reflected in the oil contracts.

Dr Salah al-Sheikhly is a softly spoken and genial economist who was Iraq's Deputy Planning Minister for most of the 1970s, until he fell foul of senior Ba'ath Party officials and went to work for the UN in New York. After Saddam took power he received a cable inviting him to return but decided against it after hearing that his cousin, a former foreign minister, had been gunned down the same day. In exile for many years, al-Sheikhly later became a senior official of the Iraqi National Accord, one of the more broadly based opposition groups. When I interviewed him in their shabby offices in Wimbledon on the eve of the Iraq invasion, he was quite explicit about the nature of the deal. The Iraqi opposition groups were being lobbied hard by American and European oil companies, and it was clear who would now have the advantage. 'Some people would be more equal than others. Obviously those who have helped us all along with regime change should have a little edge over the rest. The French may stand at a disadvantage.'[59] Dr al-Sheikhly now works in somewhat smarter accommodation, as Iraq's ambassador to London.

It would be a mistake, however, to think that the invasion was principally about profit. Although there was a general understanding with the Iraqi opposition that British and American companies would benefit, the decision to invade clearly had more strategic roots. The idea seems to have been to prise open Iraq and the wider Middle East for Western investment – not only for the sake of profit, but in order to raise oil production. The invasion was not just 'all about oil'; it was about peak oil.

For the US, we know from their public statements that Cheney and the Neocons were exercised about depletion and the replacement of dwindling Western reserves. And the leaked NSC documents make clear the administration had conflated its 'policies towards rogue states' (plural, notice) and 'the capture of new and existing oil and gas fields'. For Britain, its recent experience of the sharp onset of terminal decline in North Sea production and of a paralysing fuel crisis – although unrelated – gave every reason for extreme sensitivity about strategic oil supplies. The soothing public statements from the DTI must be set in the context of the secret creation of a permanent diplomatic liasion between Britain and America, whose purpose was to foster 'energy security'. If the government had not fully appreciated the extent of worldwide oil depletion before, it now had a bespoke hotline to an administration that certainly did, and it seems Britain was keen to exploit the relationship. A despatch to Washington from the US Embassy in London in September 2002, released to me under the US Freedom of Information Act, reported: 'The British strongly support this dialogue with the U.S. and want to use it to leverage U.S. and UK influence on energy issues in Russia, the Caspian and the Middle East . . . Officials report that the Prime Minister's office request regular updates on the preparatory work.'[60] Both governments continue to withhold the actual minutes of the meetings of the Dialogue. So while Blair may have taken Britain to war in any case, it is hard to avoid the conclusion that an awareness of the impending oil peak was part of his private thinking. Whether the invasion was a remotely sensible energy policy, even in terms of cold-blooded *realpolitik*, is of course another matter completely.

Before the invasion it was confidently predicted that Iraqi oil production would grow strongly afterwards, perhaps doubling by 2010, with the help of international oil companies. Even some of the more sober assessments predicted that production would hit 4 million barrels per day within a couple of years. [61] But it hasn't quite worked out like that. In the chaos and savagery of post-invasion Iraq, far from growing, oil production has fallen further, and no major foreign oil company has dared to set foot in the country.[62] The slow strangulation of UN sanctions has been replaced by daily acts of sabotage against pipelines and refineries. As a result Iraqi oil production languishes below even its sanctions-bound pre-invasion level, and Iraqis live with the grotesque irony of chronic shortages of fuel, for cooking, power and transport. However, none of this undermines the idea that the *intention* was to increase oil output; rather it highlights the delusions of Rumsfeld, Perle, Wolfowitz and the rest. These people truly believed the invasion was going to be a 'cakewalk' and that the Marine Corps troopers would be garlanded in Baghdad.[63] No wonder they were wrong about the oil output too.

The wider purposes of the Iraq invasion are also evident from a quick survey of events in the other oil-producing countries on America's sanctions hit list – the ones Dick Cheney wanted to 'review' in the light of his energy security concerns. Here things seem to have gone closer to the script. Colonel Gaddafi for instance took the hint rather quickly. By December 2003 he had 'renounced' weapons of mass destruction, Jack Straw called him 'courageous and statesman-like' and Libya was open for business. In early 2005 foreign oil companies were invited to bid for exploration licences, the first time any OPEC member had done this since the nationalizations of the early 1970s. Of the first tranche of fifteen licences, eleven were awarded to US firms. Libya has the biggest and highest quality reserves in Africa, and with Western investment now intends to double its production to 3 million barrels per day by 2010. It's true that a rapprochement had been developing for some time but, as the *Financial Times* reported, the invasion of Iraq was critical in forcing the capitulation: 'Some Arabs who were in regular conversations with Libyan officials say the regime was increasingly desperate to secure a deal as the war in Iraq loomed, worried that "it would be next" in some unspecified way.'[64] In other words, it worked.

So by the end of 2003 two of the countries whose sanctions Cheney had declared as irksome had been opened up with varying degrees of success. It seems that the third country, Iran, is work in progress.

Dallas

Louis Christian lives in a quiet, comfortable Dallas suburb, the kind where the close-cropped lawns and tidy driveways sprout crepe myrtle and basketball hoops in equal proportion. As I arrive for our interview one bright September morning in 2005, Louis greets me at the door of his spacious bungalow with a wide smile and firm handshake. He has white hair with academic-pattern balding and, I notice as he leads me through to his map room, the light step of a man who learned to tango in his seventies.

Although he officially retired years ago, Louis still keeps busy. A long career with Mobil, much of it spent assessing new international ventures, left him with a detailed knowledge of the geology of the Middle East, and the basis for a successful second career in exploration consultancy and map making. He had already sold maps of the region to many of the world's biggest oil companies such as Shell, ExxonMobil and Saudi Aramco, before he was approached by the United States Geological Survey about Iraq.

As Christian toiled over the Iraq maps he had no inkling that he might not be working solely for the USGS. As far as he knew, the Survey was conducting one of its periodic reviews of worldwide petroleum resources, and he was helping out on one small part of it. They needed him to do the work because, after years of political isolation, the existing maps of Iraq were inadequate. Christian's new maps would form the basis of the USGS computer model of Iraq's petroleum geology, from which the Survey concluded that 45 billion barrels of oil remained to be discovered.

Christian's suspicions were first aroused more than a year later, in the summer of 2001, when the Survey asked if he could produce some bigger maps of the Middle East. This led to an exchange of letters and telephone calls about the technical details and the financing of the work, which an official intimated would depend on funds from outside the Survey's normal budget. Louis fishes out a letter on which he scrawled 'State Department request' during one such conversation.

The main USGS budget comes under the Department of the Interior, but from time to time the Survey receives additional funds from other federal agencies that may have a specific interest in a particular project. According to one former senior USGS employee, funding from other agencies normally comes 'when they want us to pay attention to their priorities'. In this case however the USGS official who told Christian that they were waiting for approval from the 'State Department' added a curious-sounding qualification. 'It was something to the effect of "State Department or another agency" or "State Department or others in the administration", which led me to suspect that it wasn't truly the State Department at all but somebody else they didn't care to identify.'

Some months later the official sent him a handwritten note saying: 'I hope to hear from the State Dept. soon regarding your maps. Stay tuned.' But in the end the job fell through, and for the next couple of years Christian thought no more about it. Then the invasion of Iraq got him thinking again. His first reaction was political revulsion: 'I think the whole thing is disgusting and liable to be counterproductive.' Then he began to wonder if his maps of Iraq had in fact been funded by 'others in the administration'. 'It seems to me the most likely suspect would be somebody high up in the administration that has a virtually unlimited budget that they don't have to itemize. Most likely the CIA or the Pentagon.'

Tom Ahlbrandt is the USGS geologist who was project chief for the entire World Petroleum Assessment, and also specifically in charge of the Iraq appraisal. When I interviewed him in London in February 2005 he was intensely proud of the achievement, and repeatedly stressed the enormous effort it had involved. In the context of a discussion of the assessment of Iraq he volunteered that the work 'was funded by sources in Washington'. When I asked if that meant the State Department, he replied, 'No, it wasn't the State Department. I'm not really at liberty to talk about it.' And when I pushed him to explain when the funding had taken place, he said, 'A long time ago, when we commenced the studies.'

The man in charge of the USGS Iraq assessment had confirmed it was funded by 'sources in Washington', but when I made a formal Freedom of Information Act request for details of the funding, I received this apparently carefully worded reply: 'Our administrative officer and our project chief [Tom Ahlbrandt] have gone through numerous files for the team during the time period. They were unable to find any records indicating that the USGS received funds from any other government agency for work carried out in support of the World Petroleum Assessment 2000.'[65]

I laughed out loud, again. This was not a denial that the funding actually took place by any stretch of the imagination, merely a statement that there was no paperwork. My Freedom of Information Act enquiries to the CIA and a sweep of other intelligence services and government departments had also drawn a blank – but then what did I expect? The lack of a paper trail might even support Christian's hunch – that the money originated in some shadowy, non-itemized budget – but it did nothing to disprove the story.

Tom Ahlbrandt had confirmed the external funding in interview, and his reticence about the exact source strongly implied security or defence. This suspicion was confirmed to me by a former employee of the Survey on condition of anonymity. He explained that a senior, named colleague had told him that the Iraq work had indeed been part-funded by the CIA. Hearsay, I know, but in the circumstances, I am inclined to believe it. According to the source: 'It's my understanding that they contributed some money to stay on the cutting edge of where the study was going.'

A retired senior oilman who advised the CIA on Middle Eastern oil-fields during the first Gulf War, and who still retains his security clearance, offers a different interpretation. Like many in the industry the oilman has a low opinion of the USGS's resource estimates, and suggests that the CIA would have been less interested in the Survey's conclusions than its raw data. 'The CIA has a small department of experts who know more about fossil fuels than the whole US Department of Energy put together,' he explains. 'They have done detailed, in-house reservoir studies for many, many major oilfields around the world, to help the US be prepared for the day when demand exceeds supply.' In other words, the CIA may have paid up not for advance notice of the USGS's findings, but to secure direct access to Louis Christian's maps for their own use. Either way, it is hard to avoid the conclusion that the CIA was already sizing up Iraq's oil reserves in the mid-1990s.

Even after the invasion of Iraq, Louis Christian was not entirely convinced by his own suspicions. That day came late in 2004, when the USGS called again and asked if he could prepare some geological maps of Iran, and Christian found the request distinctly unsettling. 'I started to get a little paranoid. The last place I did a map of for them – somebody invaded.'

Tom Ahlbrandt describes their new work on Iran simply as a logical extension of their modelling studies, and volunteers, 'Our undiscovered resource estimates for both oil and natural gas are higher in Iran than for Iraq. So it's just a natural flow into surrounding regions.' This thought may also have occurred to US military planners, or their masters. When I ask him if other branches of the administration are interested in their Iran work he replies cryptically, 'Other people have interests.'

In January 2005, a couple of months after Louis Christian received the USGS enquiry about Iran, the American journalist Seymour Hersh reported that US special forces had been reconnoitring targets in that country since the previous summer.[66] In February, during a post-Iraq 'charm offensive' around Europe, President Bush exclaimed: 'The notion that the US is getting ready to attack Iran is ridiculous.' But after one of his smirking, arms-akimbo pregnant pauses, he added, 'Having said that, all options are on the table.'

The argument that in planning the invasion of Iraq both America and Britain were motivated by the knowledge of an impending shortage in world oil supplies seems to me compelling. But the evidence is stronger for American officials than for British, where the case is more circumstantial. It is just possible that Tony Blair did not know, despite the creation of the US-UK Energy Dialogue, that worldwide oil production would soon peak and go into terminal decline. But if so, he faces an unenviable choice for a Prime Minister desperate to salvage his reputation, similar to that posed by his evident duplicity over the intelligence. Blair claims not to have known that the infamous forty-five-minute claim, now totally debunked, referred only to battlefield weapons, which therefore posed no threat unless Iraq was attacked. This is surprising since the former Foreign Secretary, the late Robin Cook, claimed to have explained this to Blair in March 2003.[67] So either Blair did know it and lied, or he really didn't, and was at the very least incompetent. Similarly, either Blair did appreciate the critical state of worldwide oil depletion and led Britain into a cynical and misguided resource war, or he didn't, in which case he will be remembered as the Prime Minister who failed to anticipate what could be one of the most profound crises ever to face Britain, and indeed the entire world.

2

Dangerous Curves

T HE MAN WHO exposed the oil industry's dirty little secret half a
century ago was an irascible but brilliant Texan oil geologist called
Marion King Hubbert. In the BBC's tape vaults in London there exists
a short clip of him recorded in 1973, explaining the birth of his theory
in the mid-1950s. Filmed at an easel with some hand-drawn graphs,
Hubbert drawls, 'Common expression was, "It wouldn't happen in
my lifetime, and my grandchildren could worry 'bout oil. I don't
have to." '[1]

On film Hubbert is a kindly-looking old man, but according to
Professor Kenneth Deffeyes, who worked for him at the Shell research
laboratory in Houston and became a lifelong friend, he was fearsome:
'He could be spectacularly nasty.' Leaning back in his favourite
recliner at home in Princeton, in September 2005, hands clasped on
belly, Deffeyes recounts a typical incident during a lecture given at
Shell by a guest speaker, when Hubbert spotted an error, calculated
the correction in his head, and demolished the hapless lecturer from
the back of the hall. 'You didn't want to make a mistake with Hubbert
in the audience.'

Today Hubbert is largely remembered for his most famous theory,
but his interests were broad and his scientific achievements formidable. In 1937 he solved a longstanding geological mystery by
demonstrating mathematically how even the hardest rocks in the
earth's crust would flow like soft clays if subjected to immense pressure. In 1959 he helped solve another stubborn puzzle, explaining
the mechanism behind the displacement of huge blocks of rock,
known to geologists as overthrust faults.[2] He was elected to the
National Academy of Sciences in 1955, and in his later years became
a professor at both Stanford and Berkeley.

In 1956 Hubbert did something even more significant, by publishing a new method for predicting when oil production in any given region would reach a peak, and then slip into terminal decline. The paper was based on ideas that he had evidently been worrying away at since the 1930s, but it was not until the mid-1950s that he felt he had the data to launch his theory on the world.[3]

The story of Hubbert and his 1956 paper is usually told like this: 1) in 1956 he predicted that US oil production would begin to fall in 1970; 2) it did; and 3) the end of the world is nigh. But this is a truncated parody that completely fails to reflect Hubbert's real achievement. The story of how he developed his forecasting techniques during a decade-long running battle with his detractors illuminates a great deal about attitudes towards oil depletion in the industry and the geological establishment, and remains instructive today.

The paper was to be delivered at a meeting of the American Petroleum Institute in San Antonio. When Hubbert and his wife arrived at the hotel on the eve of the conference, their car laden with 500 copies of his paper to distribute, they were greeted by a foretaste of the controversy to come. To their surprise the couple were mobbed by reporters from the oil and gas trade press, demanding to know if Hubbert was still going to deliver his paper. 'Why, certainly,' he replied, while privately fuming; it was obvious something was up, but he didn't know what.[4]

The next morning, that something became clearer. Hubbert was to be the first speaker, but even as he was being introduced he was called from the podium to take an urgent telephone call from a Shell public relations man in New York, claiming his predictions were ridiculous and pleading with him to tone down his remarks. The PR hadn't read the paper himself, but evidently his boss had, and was climbing the walls. But Hubbert was not a man to be browbeaten or cajoled. He put down the phone and gave the talk exactly as written.

So what was all the fuss about? Hubbert's paper was called 'Nuclear Energy and the Fossil Fuels', and sought to answer the question of how humanity in general, and America in particular, could continue to satisfy a demand for energy that was growing at a ferocious pace.[5] The paper surveyed the relative size of the energy resources of coal, oil, gas and uranium, and concluded that in nuclear power the world 'at last has discovered a source of energy adequate for its needs for at least the next few centuries'. But it wasn't this conclusion that was so shocking in 1956 so much as the idea, buried in the paper, that oil and gas were going to start running short very much sooner than anybody had anticipated. At this point, nobody else in the congenitally optimistic oil industry was remotely considering the idea of shortage.

Hubbert observed that since it had taken 500 million years to accumulate the fossil fuels present in the modern world, supplies were essentially finite. True, the remains of plants and animals were still accumulating on the sea floor, some of which might eventually be buried deep enough underground by geological processes to be 'cooked' into hydrocarbons, but you wouldn't want to hold your breath. 'Therefore,' he concluded drily, 'we can assume with complete assurance that the industrial exploitation of the fossil fuels will consist of the progressive exhaustion of an initially fixed supply to which there will be no significant additions during the period of our interest.'

But consumption of fossil fuels was not proceeding so glacially. Hubbert's paper included a series of historical graphs of the production of coal, oil and gas for various regions which shared a strong family resemblance. The growth they all displayed – with production gradually picking up pace during the second half of the nineteenth century, and then heading skyward in the twentieth – is known as *exponential*.

Exponential growth can produce some really quite pleasant results. Leave £100 untouched in a bank account paying 10 per cent, and after one year it will have grown to £110, but after two it will total £121. That's because interest has been paid on the previous year's interest; compound interest is a classic form of exponential growth. As long as the interest rate stays at 10 per cent, the money will double in a little over seven years, and keep on doubling every seven years thereafter. The growth in absolute terms – the cash rather than the percentage – accelerates massively as the doublings continue. After fifty years the account will swell to £11,739, after a hundred years whoever inherits it will be a millionaire, and after another hundred years a billionaire.

But the results of exponential doubling are not always so agreeable, a point well illustrated by a famous Persian myth.[6] A cunning courtier gives the king an ornate chessboard, for which the king feels obliged to offer something in return. The courtier suggests an apparently modest deal, to which the king (whose maths was evidently as good as mine) agrees: one grain of rice on the first square, two on the second, four on the third, and so on. By the twentieth square the king owed more than half a million grains, and by the sixty-fourth the figure had grown to 18,446,744,073,709,551,616 grains, more rice than existed in the entire world. History does not relate whether the courtier kept his head, but he certainly didn't receive payment in full.

These examples neatly illustrate the basic problem with exponential growth: there are not many things in life that can keep it up for long. Money, which is man-made, virtual, and effectively limitless, may theoretically grow exponentially for ever, but things like rice and oil which are finite clearly cannot.

Hubbert's steepest graph was for US oil production (figure 2), which in the years between 1880 and 1930 grew at almost 8 per cent a year. This meant that output doubled in less than every nine years. If it carried on at this rate, he reasoned, it would not be long before output reached impossibly astronomical levels. 'The world will only tolerate so many doublings of anything,' he said later, 'whether it's power plants or grasshoppers.'

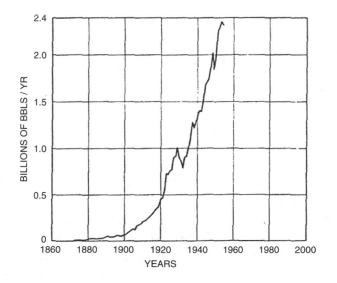

FIGURE 2. US oil production to 1956[7]

But Hubbert had also spotted something else: that while oil production was still growing strongly, the rate of growth was in fact slowing down. This suggested to Hubbert that output was being inhibited by natural physical limits. If that was right, he figured, it would continue to slow until it stopped growing altogether, and then decline exponentially back to zero. As a result the graph would look something like a bell-shaped curve, a shape now known in the oil business as Hubbert's peak.

To find out when the peak would occur you simply needed a good estimate of the total oil ever likely to be produced in the US, known in modern jargon as the *ultimate reserves* or, in the oilman's shorthand, *ultimate*. (This is not the total amount of oil that exists, but the total that the industry will ever manage to extract, which is a much smaller number.) The figure for ultimate would have to include all production to date, known reserves still in the ground, and an estimate of what was likely to be discovered in future. At the time the best estimates for ultimate ranged from 150 billion barrels to 200 billion barrels, so Hubbert plotted a curve for each (figure 3). There was no fancy maths involved, and as this was long before the days of computers, he drew them by hand.

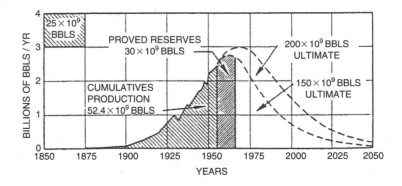

FIGURE 3. Hubbert's famous graph used two estimates of US ultimate, the total amount of oil that would ever be produced: 150 billion barrels, and 200 billion barrels. Both figures subsume the amount of oil produced up until 1956 (the hatched area), and the known underground reserves at that time (the finely hatched area). The white area under the two dashed lines represents the amount yet to be discovered in each estimate[8]

Since the first seventy years of production (1880–1950) were known, history had already drawn much of Hubbert's graph for him, and the total area under the curve was absolutely fixed in each case by the estimates of ultimate. Given these constraints Hubbert found there were only a very limited number of ways he could plot a plausible-looking curve while at the same time satisfying his self-imposed conditions of incorporating the existing production history, sticking to the estimates of ultimate, and returning the graph line to zero. The results were startling: for the lower estimate it turned out that peak production would be reached just nine years later, in 1965; for the higher estimate the date was 'about 1970'. Either way, Hubbert declared, US oil production was going to go into terminal decline within fifteen years. *That* was what all the fuss was about.

Although he never quite says so explicitly, from Hubbert's graphs it's clear that oil production starts to fall when only about half the oil has been produced, *and half is still underground.* To anyone who is not in the oil business, this may seem an extraordinary idea; it is not immediately obvious why production cannot continue to rise until the last drop comes out of the pump. But Hubbert's notion that oil production in a given region or country must start to fall at the midpoint of depletion or thereabouts was based on factors with which his audience of oilmen would have been thoroughly familiar. The 'natural limits' to which he alluded were limits that petroleum engineers battle with every day.

Those limits derive from two crucial facts of life in the oil business, and the first is to do with pressure. Most people don't realize that a reservoir is not an underground lake, but made of solid rock, or almost solid. The oil and gas are trapped in tiny interconnected pores, from which you might think they would be impossible to extract. But luckily the reservoir usually exists under great pressure, and a typical deposit 10,000 feet below ground might be under about 4,500 pounds per square inch (psi). After the well has been drilled and the necessary plumbing installed, the differential between that pressure and the much lower pressure at the surface (just 15psi) forces the oil up the pipe of its own accord. So to start with, nature does all the heavy lifting.

But this serendipity does not last for ever, and sometimes fades very quickly. From the moment the oil starts to flow, the pressure in the reservoir begins to fall, and so does the well's maximum possible rate of production. Every week, month or year the well will produce less oil than the one before. At some point (about 2,500psi in our example, if the oil were about average density) the pressure in the reservoir becomes too weak to lift the 10,000-foot column of liquid, and the oil simply stops flowing altogether.

At this point the oil company has a choice: is it worth putting a lot of effort into getting some more of the oil out by artificial means? If the costs can be justified, the company might drill another well some distance away and pump water or natural gas down into the reservoir to maintain the pressure, and so keep the oil flowing a bit longer. But then comes the next problem. When the water or gas reaches the intake of the production well, as eventually it must, it displaces much of the oil going up the pipe, so the oil production rate suddenly falls dramatically. The company can also lower an electrical pump to the bottom of the well to help push the oil up, but the impact of such 'artificial lift' is limited. In fact the whole process is a battle to maintain flow rates, in which the company is the inevitable loser from the very start. As the field gets older and wearier, each new attempt to stem the decline costs more and yields less, and eventually the company packs up and moves on. On average this happens when they have recovered just 35–40 per cent of the oil in the reservoir. That's right: almost-two thirds is usually left behind.

Falling pressure and rising water intake mean that the natural ability of an individual well to produce oil will start to fall from the first day of production. But since most fields have more than one well, and sometimes several hundred, production tends to rise sharply to start with as more wells are brought on stream. For big fields this growth spurt may last for years, for smaller ones it could be just a few months. In either case, once the addition of new wells stops, the falling pressure takes over and decline begins. Production profiles for individual fields vary greatly, but the Forties field (figure 4) in the British North Sea is fairly typical. Barely five years of growth have been followed by twenty-five years of decline, giving a peak skewed significantly to the left.

The only way an oil company can avoid this decline for any period of time is by deliberately choking off the flow of oil to below

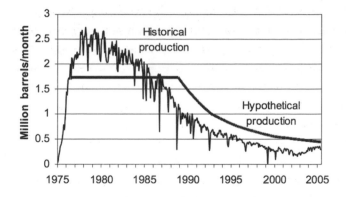

FIGURE 4. The Forties field shows a classic leftward-skewed production profile, with output falling from 2.75 million barrels per month in 1979 to less than half a million today. (Source: DTI) The smooth line shows how a deliberately restricted flow rate could be held constant for a time, but that decline reasserts itself eventually

maximum capacity. If the operator tightens the spigots by a couple of turns, the oil will flow steadily at the reduced rate for a time. On the graph this would look more like a plateau than a peak (figure 4, smooth line). About the same total amount of oil may be produced in either case, but the plateau has the effect of delaying the production of a portion of the oil until later.

Ordinarily an oil company would want to avoid doing this; the business school doctrine of 'net present value' means that oil sold today is always worth more than oil sold tomorrow. But in some circumstances the costs of producing at full blast, requiring more wells and bigger pipelines, may be prohibitive, and it may make sense to economize on the infrastructure and produce more slowly over a longer period. In this case, the production profile appears to defy gravity for a time, but this masks the fact that the production *capacity* of the field is still declining all the while. When the field's production capacity falls below the rate at which the oil is actually being produced, decline takes over again. This kind of production profile is characteristic of gas fields.

In these fairly typical examples, nature and the oil industry seem to have conspired to produce either no peak at all, or a peak in the wrong place. If Hubbert was right, shouldn't there always be a peak more or less in the middle? The discrepancy is easily explained, however, since this is a graph of a *single field*, and the Hubbert peak relates to an oil-producing region or country, containing perhaps thousands of fields. When you add up the production from all the fields, it produces an interesting and ultimately terrifying effect.

This is caused by the second crucial oil business fact of life. When an oil company sets out to explore virgin territory (not much of that left, by the way), they almost always find the biggest fields first. Even if the geologists never left the warmth of their Portakabin, threw darts at a map and drilled at random, probability tells you they would hit the so-called 'elephants' early on. These days exploration is a bit more sophisticated than that. Seismic surveys have been in use since the 1930s, and today superior three-dimensional seismic is used routinely, helping geologists to identify even the smallest, subtlest and deepest traps that might contain oil and gas. This technical sophistication combines with the profit motive to reinforce the effects of probability: the largest fields are discovered early, and the finds become progressively smaller until the whole region is well understood and drilled out.

It is these factors that combine to explain the reason why the Hubbert peak exists in nature, through an effect beautifully illustrated by a simple theoretical model devised by Roger Bentley of Reading University (figure 5). This shows graphically why individual field production profiles that are skewed to the left can add up into a much more central peak for their region as a whole. The model assumes only that one new field is brought on stream per year, and that each is 20 per cent smaller than the last. Look what happens!

The resulting peaks are not as symmetrical as those of Hubbert's 1956 paper, but then Hubbert himself stressed that there was no theoretical necessity for this; it was just the way it came out when he drew the curves for America. What is clearly shown here is how falling production rates combine with falling field sizes to create a Hubbert-style peak and decline. This is the third oil business fact of life, so get used to it.

Hubbert's paper caused uproar in the industry. When he returned to Houston some days after delivering it he found the mood in the Shell office extremely tense. 'Apparently all hell had been going on in my

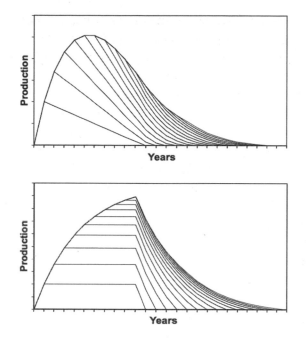

FIGURE 5. The Bentley Curves. In these models it is assumed that one field starts producing each year, and that each is 20 per cent smaller than the last. Regardless of whether the individual production profiles are left-skewed peaks, or a series of plateaux, more common in gas production, the size distribution of the fields still produces an overall peak at about the mid-point[9]

absence.' It is easy to see why. Nobody likes to be told the party will soon be over, and most could not believe it.

At the time the US oil industry's stock answer to questions about shortage was that America had all the oil it needed for the foreseeable future, which Hubbert described as a 'beautiful case of prediction by ambiguous statement'. This complacency would echo down the decades: in the face of all evidence to the contrary, British ministers and bureaucrats today use *exactly* the same phrase to describe the position of worldwide oil supplies. As a Foreign Office official wrote to me in 2005: 'Our economic advisers consider that there should be enough oil for everybody's needs for the foreseeable future.'[10]

In 1956 America such a view, although unfounded, was understandable. The US had been in the oil business for a hundred years and produced about 50 billion barrels. If the estimates of ultimate that Hubbert had employed for his forecasts were right, America had only produced between a quarter and a third of its total. In other words, there remained still two or three times as much again to produce. So the intuitive judgement of most oilmen was that there could be no short-age in their lifetime. Hubbert's paper 'jolted the hell out of them'.

Some were not only incredulous but actively hostile. Humble, a subsidiary of Standard Oil of New Jersey, which would later become Exxon, was the largest oil operating company in America at the time, and so had the most to lose if production started to fall. A senior executive named Morgan Davis, whose career at the company spanned Chief Geologist and chairman of the board, took particular exception to Hubbert's ideas and attacked them regularly at industry conferences. This running battle continued for over a year, and evidently became quite personal in the Davis household. At an industry lunch Hubbert found himself sitting next to Davis's wife, and they were chatting sociably, although they hadn't actually been introduced. When another diner addressed Hubbert by name, 'Mrs Davis froze visibly.'

Mr Davis was apparently determined to prove Hubbert wrong by any means necessary. According to Hubbert, the Humble executive would always 'come up with great nebulous amounts of oil and gas, but never anything you could hang him with'. Finally in November 1958 Davis came up with an ingenious idea. He proposed changing the definition of reserves in a way that, coincidentally, would make them appear much larger. With a stroke of the pen, the impending peak and decline in American oil production would disappear, or at least fade comfortably far into the future. Of course Davis himself did not admit this was his motive, but Hubbert was convinced. 'He was jimmying the basis of prediction . . . Then you won't upset your applecart.'

This technique seemed to catch on, even apparently at the once revered United States Geological Survey, the official body which is meant to provide the American government with authoritative assessments of the country's natural resources. The USGS was for years regarded as the acme of geological study in America, *the* setter of standards, *the* place to work. There was a time when American geology undergraduates were made to learn by heart the names of all the directors of the agency since it was founded in 1879. But since Hubbert the USGS has been accused of producing absurdly high estimates of natural resources, which its critics argue are the result of a perverse methodology devised by a politically influenced bureaucracy.

In the years following Hubbert's bombshell, various estimates of US ultimate were published which leapfrogged each other – 250, 300, 400 – until in 1961 the USGS trumped them all with the astonishing figure of 590 billion barrels, three times greater than the larger of the two industry estimates Hubbert had used. Hubbert suspected these estimates were deliberately concocted to discredit his predictions, and later joked that his 1956 paper had generated more oil than the entire industry had discovered in a century of exploration.

The USGS figure was based on the work of a geologist named A.D. Zapp, evidently a serious and scholarly scientist, whose reputation may have been unfairly treated by posterity; he died soon after the figures were published and had no chance to modify them. Zapp's work was seized on by Vincent McKelvey, then the assistant chief geologist at the USGS, and later its director, who wrote: 'Those who have studied Zapp's method are much impressed with it, and we in the Geological Survey have much confidence in his estimates.'[11] These estimates may have been just what McKelvey was looking for; he would use them as ammunition in a running battle with Hubbert lasting well over a decade.

The cause of McKelvey's evident hostility towards Hubbert is not entirely clear, but it may stem from events to do with the preparation of a blue ribbon report by the National Academy of Sciences for President Kennedy on the state of America's natural resources. As an eminent member of the Academy, Hubbert was chosen to chair the energy section of the report, and by Hubbert's account McKelvey was 'mad as hell' not to have got the job himself. There may be another reason less to do with personal ambition and more to do with the politics of a large bureaucracy. Would the USGS have more status, budget and interesting work if it admitted that oil was a dwindling resource, or if it claimed there was lots left to find? Might accepting Hubbert's conclusions have directly threatened the interests of the organization as well as the egos within it?

Perhaps by no coincidence, it was during the preparation of the Academy report that the USGS came up with its figure of 590 billion barrels. However absurd Hubbert considered that number, it did highlight a weakness of his technique: that it depended on an estimate of ultimate at a stage when nobody could be certain what the number would, eventually, turn out to be. What if the estimates Hubbert had used in 1956 were indeed wrong? Although he did not believe they were, for a rigorous scientist like Hubbert this thought rankled. So for the Academy report he went back to the blackboard and worked out another entirely new method of predicting the peak of production, one that would dispense with the need for an ultimate reserves figure altogether.

Hubbert had noticed for some time that the graph of US cumulative production (the total oil produced so far) had mimicked the shape of cumulative discovery rather closely, with a time lag of a little over ten years. From figure 6, it is clear that cumulative discovery reached 40 billion barrels by about 1940, and that cumulative production caught up to that level by about 1950. This makes intuitive sense: you can only pump oil that has already been discovered, and *if* the average time taken to lay pipelines, drill production wells and so on remains about constant, production should follow the shape of discovery reasonably closely.

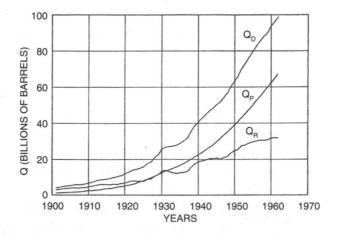

FIGURE 6. Hubbert's second technique. Cumulative discovery (top line) reached 40 billion barrels by about 1940, and cumulative production (middle line) caught up to that level by about 1950. If the gap held constant, the production peak would come about ten years after the discovery peak[12]

Hubbert had also spotted that since 1956 the annual rate of oil discovery had been falling. America was still finding new oil, so the cumulative graph line would continue to rise, but each year less was being discovered. In figure 6 you can just about see the 1956 turning point, where the cumulative discovery line begins to ease off, just to the right of the letters *Qd*. On a graph showing the *rate* of discovery this would be the peak.

American discovery was falling for same reasons it does in any mature oil-producing region, although being the first major province to pass its discovery peak this was little understood at the time. Hubbert was the only one to work out what it really meant. He realized that because discovery had peaked in 1956 it followed that, if the ten-and-a-half-year time lag held good, production would peak halfway through 1967. This was also about midway between his previous predictions for the peak of production, so the new technique agreed with and reinforced the results of Hubbert's original paper.

This much Hubbert could have deduced intuitively simply by looking at the historical data in figure 6. But the graph lines also suggested to him a similarity with the *logistic* curve, devised by the Belgian mathematician Pierre-François Verhulst in 1838, and this insight allowed him to make another breakthrough. There is no rule that says oil production must follow the pattern set by the logistic curve, but the American historical data seemed to fit, so Hubbert set to work to find out what this could tell him.

The logistic curve is traditionally the shape of a flattened-out 'S' on a graph, and was invented to describe population growth – fish in a pond, or weeds in a field – and to predict how that growth will be constrained by the carrying capacity of a given environment. It may seem odd that this has anything to tell us about oil production, but there is a connection. In biology, population growth tails off as the size of the population gets closer to its carrying capacity. Similarly, Hubbert's work was predicated on the obvious truth that oil is harder to find the smaller the fraction that remains to be found. These are variations on the same theme: the smaller the headroom, the lower the growth. So the logistic curve might be relevant to oil production.

Now for the really clever bit. If you have enough historical data to plot the early part of the curve, the logistic equation allows you to predict its future path and deduce the maximum population that will ever be reached, long before it has actually happened. Hubbert applied this to the US oil discovery and production data, and the maths allowed him to derive a figure for ultimate, the total that would ever be produced in America. By solving the logistic equation using historical discovery and production data (figure 7), Hubbert was able to plot the future path of both, and found that they flattened off at about 170 billion barrels. Again, this figure fell about midway between the two estimates he had used in 1956. So Hubbert's 1962 paper not only did away with the need for an initial estimate of the total oil that would ever be produced in order to predict the date of peak production, it also came up with a brand-new way of calculating – rather than guesstimating – ultimate, which reinforced his earlier work.

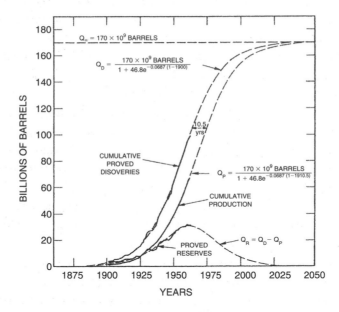

FIGURE 7. Hubbert used the logistic equation with historical data for discovery and production to calculate the total amount of oil that would ever be produced in the US[13]

As alarming as it was, Hubbert's new forecast had absolutely zero effect on US government policy. His energy report was buried in a tome that covered all of America's natural resources and was delivered to President Kennedy in December 1962. Politically the timing could hardly have been worse. Hubbert was predicting US oil production would peak at least by the end of the decade, yet Kennedy had just promised to put a man on the moon within the same timeframe. Hubbert's was not the kind of optimistic message America wanted to hear. Besides, the President had a few other things on his mind: America was just beginning to get seriously entangled in Vietnam, and the world was still palpitating after the Cuban missile crisis. It's hardly surprising the 1962 report sank with barely a ripple.

Vincent McKelvey must have been delighted. Hubbert had rejected his USGS estimates for the Academy report 'because they could not remotely be reconciled with the petroleum industry data',[14] and now the report had been buried. Not only that, but US oil production continued to rise, demolishing Hubbert's two nearest-term predictions of peak production: 1965 and 1967. From the raw production figures there was nothing to suggest oil output was about to go into terminal decline. Emboldened, the USGS continued to publish their 590 billion barrel figure for the rest of the decade, and McKelvey was promoted to Chief Geologist.

In 1964 Hubbert retired from Shell and took up both a professorship at Stanford and a position at the USGS, where McKelvey was now his boss. Perhaps by no coincidence, Hubbert arrived to find that it had been arranged that his new job would have nothing to do with the contentious issue of oil reserves. Nevertheless he was still determined to get to the bottom of why the agency's figures were so high. He had always believed they were absurd, but it was only after 'stewing' over them for several years that he worked out what was wrong.

The USGS estimates were based on the 'Zapp hypothesis', which itself worked on fundamentally different assumptions from Hubbert's own. Instead of interpreting data generated by real-world oil industry behaviour, such as the rates of discovery and production, Zapp took an essentially *volumetric* approach, by trying to calculate how much potentially oil-bearing rock existed in theory, and how much oil it might yield.

According to an industry estimate at the time, America contained about 1.8 million square miles of land with the necessary sedimentary rocks favourable for oil discovery. Zapp calculated that to find all the oil would mean drilling 5 billion feet of oil wells.[15] Zapp also calculated that so far the industry had drilled about 1.1 billion feet and discovered about 130 billion barrels, an average of 118 barrels per foot. So it followed that if they went the full nine yards, or rather 5 billion feet, and if their average success rate stayed the same, they would discover 590 billion barrels. That was a very big *if*.

Hubbert pounced. Zapp had taken only *one* drilling success rate average for the entire history of the US oil industry (1859–1960), and extended it into the future. But how could this be right? Hubbert reasoned that because the biggest fields, and those nearest the surface, were generally found first, large amounts of oil would be discovered early on with relatively little drilling. As the fields remaining to be discovered became smaller and deeper, it would take more drilling to find them. The drilling success rate ought to fall.

In his 1967 paper, Hubbert divided the total drilling done so far into columns, each representing 100 million feet of drilling effort, and calculated how much oil was discovered in each interval.[16] Figure 8 shows that for the first 100 million feet (1859–1920), 194 barrels of oil were discovered for every foot drilled. In column three (1929–37) discovery soared, despite the Great Depression, simply because the vast East Texas field was found in 1930. After that, the discovery rate collapsed roughly exponentially to just 35 barrels per foot by the mid-1960s.

This insight not only convincingly nailed the absurd Zapp/USGS estimate, but almost in passing gave Hubbert yet another method to calculate ultimate. Hubbert worked out the negative exponential decline that best fitted the barrels-per-foot data and extended that decline rate into the future. This gave him an ultimate figure of 168 billion barrels, which Hubbert thought was a bit on the low side, but which nevertheless once again chimed with his earlier forecasts. On this basis he concluded that peak production 'will probably occur near the end of the 1960 decade, and the ultimate amount of oil production will probably be less than 200 billion barrels'. This was an intellectually brave forecast, since the witching hour was barely three years away, two of his predicted peak dates had already come and gone, and production was still rising.

By this time Zapp was long dead, and had no chance to modify his work. McKelvey was still alive however, and even though the basis of the USGS estimate had been demolished, he continued to promote

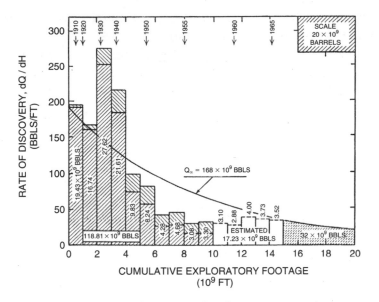

FIGURE 8. Hubbert's 1967 technique. Each column represents the amount of oil discovered per 100 million feet of drilling, which shows a roughly exponential decline. From this Hubbert deduced another method of calculating ultimate[17]

it. By now Hubbert was convinced the USGS Chief Geologist's motives were malign. 'Up until the mid-sixties I think McKelvey really believed his own hallucinations, but there was no basis for it thereafter, and all he was really out to prove was that he was right the first time,' Hubbert later recalled. 'I could only interpret this as inexcusable dishonesty.'

Hubbert had trounced his opponents intellectually, but still his ideas failed to make any wider impact in government or the industry. Resource optimism was still the oppressive orthodoxy. Even as production peaked in 1970, right on schedule, executives slept soundly in the comforting knowledge that the US was producing more oil than it ever had. Of course it was. It would take several years and an international crisis for the American decline to become apparent. McKelvey meanwhile must have laughed himself hoarse; he was promoted to director of the USGS. It wouldn't last long.

The first real evidence that American oil production had slipped into terminal decline came in the spring of 1971, when an obscure notice in the *San Francisco Chronicle* read: 'The Texas Railroad

Commission announced a 100 percent allowable for next month.'[18] Not many who even spotted this would have understood. Oil used to be transported largely by railway tankers, so in the days when Texas had spare production capacity, the Commission got the job of rationing production, making sure the Lonestar State didn't flood the market, drive the price down and everybody else out of business. The press notice reflected the fact that the industry could pump flat out but still not meet demand. The Texas Railroad Commission was now effectively out of a job, although its spirit would live on in OPEC.

The Organization of Petroleum Exporting Countries (OPEC) was set up in 1960 to secure a better deal for its largely Middle Eastern members from the international oil companies, which until then had imperiously dictated terms. The man principally responsible for its creation was the Venezuelan oil minister Juan Pablo Perez Alfonso, who had studied the operations of the Texas Railroad Commission during an earlier period of exile in the United States. But in its early years OPEC enjoyed neither the discipline nor the market conditions to allow it to exert that kind of power. It was only in 1973, after the decline in US oil production was well established, that the cartel could seize control of the world oil market by slashing its output and forcing the price up fourfold. This was absolutely no coincidence.

OPEC had tried using the 'oil weapon' before, during the Six Day War with Israel in June 1967, but with little success. Within days of the outbreak of war, Arab producers had cut their exports by 1.5 million barrels per day to punish Israel's allies – America, Britain and West Germany – but it hardly mattered. At this stage America still had 'shut in' or spare production capacity: fields that had already been developed with the necessary wells and pipelines, but were not being used at that point. Simply by opening the spigots, the US was quickly able to increase its production by 1 million barrels per day. Venezuela and Iran, both then friendly towards America, also raised their output and the OPEC challenge was defeated.[19]

By 1973 everything had changed. Falling American oil production and rising demand meant the US had run out of spare capacity. In the preceding three years alone its imports had doubled to more than 6 million barrels per day. On the other side of the equation, Saudi Arabia's share of world oil exports had soared from 13 to 21 per cent.

So when Egypt and Syria launched the Yom Kippur War against Israel in October 1973, and OPEC imposed what became known as the 'Arab oil embargo', America was powerless to raise its oil production. Quite the contrary. US output dropped from 4.26 billion barrels in 1973 to 4.1 billion the year after, and just 3.96 billion the year after that.[20] So despite the most extreme economic and political incentives to produce more, American output continued to fall. In other words, the first oil shock made it excruciatingly obvious that Hubbert had been right all along.

Finally Hubbert began to be taken seriously. Stewart Udall, who had been Secretary of the Interior under Kennedy and Johnson for most of the 1960s, and before that had written an environmental best-seller *The Quiet Crisis*, became his champion. 'Hubbert was a very important, prophetic figure,' says Udall, now in his eighties and with failing sight, but whose memory of those years is still sharp.

During his time in office Udall had believed the USGS, which came under his department, was 'one of the very best scientific organizations in government', and was not aware that Hubbert had destroyed the credibility of its oil resource estimates. (I don't suppose that's how the USGS reported it upwards.) So the arrival of the first oil shock forced him to revise his view of both the agency and its director. 'I realized the nation during my period in office had been misled. People like McKelvey were saying "Oh there's vast supplies that haven't been discovered." They loved the word "vast" in describing the undiscovered oil . . . I was appalled that a scientist – a trained scientist – could take the attitude he took with something as vital as this was to the United States of America.'[21]

Once the new facts of life were fully appreciated, Vincent McKelvey's position became increasingly precarious, especially when a group of his own scientists finally came up with a set of numbers closer to Hubbert's – and reality.[22] Finally, in 1977, McKelvey was sacked. It is no surprise that the official history of the USGS contains barely a word about this whole episode, and today the man himself is little remembered. There is a mountain in Antarctica named after McKelvey, but perhaps his lasting legacy will be that he provoked some of Hubbert's finest work.

The short version of the Hubbert story elevates him to the status

of folk hero by concentrating on the fact that one of his earliest predictions of the US oil production peak hit the bull's-eye. But this is
to miss the broader point. Because it relied on industry estimates of
ultimate, the accuracy of the 1970 prediction may have involved an
element of luck. Hubbert's real achievement was to have developed a
number of different techniques that produced a series of predictions
that were all very closely clustered and broadly right: 1965, 1967,
1969, 1970. At a time when nobody else in the industry was even
thinking about when American oil production would go into terminal decline, and much of the geological establishment was actively
hostile to the very idea, *all* Hubbert's predictions were astonishingly
accurate, and any one of them should have been sufficient to earn the
respect of the industry and the wider world.

Of course Hubbert's fundamental legacy was the idea of 'peak'
itself, and his insight that this would occur at about the halfway point.
This came with a terrifying implication: if it was true for individual
oil-producing regions like the United States, one day it would also be
true for the world as a whole. The only question was when.

3

The Wrong Kind of Shortage

T HE WORLD FINALLY started to wake up to the precariousness of its oil supply the morning after Katrina howled into New Orleans at the end of August 2005. Along with the human misery it visited on the city's inhabitants, the hurricane also destroyed billions of dollars in vital oil and gas infrastructure in the Gulf of Mexico. The wind and waves tossed drilling rigs about like toys in a bathtub, fractured underwater pipelines, and peeled the roofs from refinery storage tanks like so many tin cans. A single storm promised to shut down production of about 1.5 million barrels a day for months to come, and pushed the oil price to a new record high of more than $70 a barrel.

The market was only restored to some kind of fragile balance by the release of emergency stocks of crude and petrol by the International Energy Agency (IEA), for just the second time in its history. The IEA had been set up after the first oil shock in 1973 by the Organization for Economic Cooperation and Development (OECD) to co-ordinate the response of the industrialized countries to any subsequent energy crises, but its emergency plans had almost always stayed in their drawer. Far bigger shortages than Katrina had been weathered over the years without drawing on member countries' emergency stocks.[1] But now the IEA judged that even though the Katrina shortfall was smaller, the threat was greater. Something was up.

By chance my research trip to Houston was sandwiched between Katrina and Rita. The arrival of the second hurricane a few days later would casually capsize a massive Chevron production platform – ironically named *Typhoon* – and turn Interstate 45 into a fifty-mile parking lot, an event now known to the locals as the 'Texodus'. Rita would also force President Bush into an extraordinary about-face on energy policy. As supply problems started to bite, he urged Americans to cut out unnecessary driving and declared, 'We can all pitch in by being better conservers of energy.' Since Dick Cheney had accurately characterized the thrust of Republican energy policy in 2001 with the words, 'Conservation may be a sign of personal virtue, but it cannot be the basis of a sound energy policy,'[2] this had to be serious.

Already in mid-September the motels were booked up with disconsolate refugees from New Orleans, the gas stations full of motorists grumbling about the price of fuel that had doubled in a year to $3 a gallon. Among the energy cognoscenti I had come to visit, the discussion was all about the latest damage reports, and about what precisely was in short supply.

The IEA's emergency action inadvertently highlighted one of the more immediate shortages. The plan was that OECD member countries would offer to release 2 million barrels a day from their emergency stocks, of which half was to come from US stores. But America only holds emergency stocks of crude, and now there were fewer refineries available to turn it into anything vaguely useful like petrol or jet fuel. European countries hold some of their emergency stocks in the form of refined products, so twenty-five tankers with gasoline cargoes were immediately diverted from Europe, raising petrol prices there by 30 per cent.[3] European drivers shared America's pain.

With almost 30 per cent of its refinery capacity shut down by the storm, there was obviously no spare capacity in America, but it turns out there was none anywhere else in the world either. By the end of 2004 the pipework of the world's refineries had the capacity to process 84.6 million barrels a day, but during 2005 average demand was 83.5 million barrels a day. So on the eve of the hurricanes, refiners were already effectively working flat out and, for the first time anybody could remember, actually making money. The day after Katrina, any Gulf coast refinery still left standing saw its margins triple.[4]

Since untreated crude is no use to anybody, the destruction of refinery capacity wrought by Katrina and Rita created – or tightened – a serious bottleneck. But on the face of it there should have been no shortage of actual crude oil, at least not globally. Even with all Gulf of Mexico production shut down, the hurricanes took out 1.5 million barrels a day at most. OPEC was thought to have about

2 million barrels a day of spare *production* capacity – that is, oilfields that were already fully developed with the necessary wells and pipelines but not currently being used – mostly in Saudi Arabia. All the Saudis had to do was turn the spigots and the oil would be available.

Except the problem was that it was the wrong sort of oil. The crude that Saudi could offer was not the 'light sweet' variety favoured by the industry, but mostly 'sour', meaning it contains a lot of sulphur, and 'heavy'. Both characteristics demand specialized refineries to remove the sulphur and lighten the oil, and there wasn't enough of this kind of refinery capacity. So the world's 'spare' production capacity was meaningless; in effect there was none. Sitting in his fiftieth-floor eyrie in downtown Houston, energy banker Matt Simmons recalled, 'We ran out of everything.'

The fact that the world was running out of any meaningful spare production capacity had been obvious within the oil business for more than a year. The Oil and Money conference is one of the industry's big annual bashes, and at its week-long meeting at London's Inter-Continental Hotel in October 2004, the buzz among the delegates was about how unexpectedly strong demand growth had left oil producers struggling to keep up. Much of the growth came from the emerging economies, especially China. Demand in the People's Republic had grown by an astonishing 1.5 million barrels a day in a single year, the equivalent of three-quarters of Britain's North Sea production. To almost everyone's surprise, the industry was now pumping just about flat out, and seemed to be scraping up against some kind of production ceiling. The critical question that informed all the debates that week had to do with the nature of that ceiling: was it just the result of past under-investment, that would soon be remedied now the oil price had surged to $50; or was it the first inkling of a fundamental geological constraint that would prove immovable? Was it just about the plumbing above ground, or did it in fact represent the first tremors of the last oil shock?

At the Oil and Money conference, the 'official' view from the podium was relentlessly optimistic. The IEA used the conference to launch its biennial energy markets analysis, the *World Energy Outlook*, a fat red book that forecast that world oil production would in fact grow by 40 million barrels a day to an astonishing 120 million barrels a day by 2030. IEA chief economist Fatih Birol told the conference that the apparent capacity ceiling was purely financial: 'Let me turn to the question of the day, whether or not we are running out of oil. Our answer is a definite no. We have enough oil and gas reserves up to 2030 and beyond to meet our demand, if the investments are done at the right time and in the right places.' The keynote speaker, BP chief executive Lord Browne of Madingley, was bolder still. In a talk entitled 'Beyond Insecurity – The Future of the World Oil Market', he declared that despite appearances to the contrary, 'we have to demonstrate that there has been no shortage of oil, and that there is no shortage of oil, and that there never need be a shortage . . . there is no reason why there should be any shortfall in the foreseeable future.'

You could be excused for thinking that two such widely respected authorities ought to know what they're talking about. The IEA is the energy co-ordinating agency for the OECD, and its analysis is relied upon by Britain and many other countries to make policy. Lord Browne is arguably the most successful oilman of his generation: after gaining a first in physics at Cambridge, and joining BP as a petroleum engineer, John Browne claimed the top job when he was only forty-six. The company was still recovering from a financial crisis in which it had been forced to halve its dividend, but within a decade Browne had transformed it into Europe's largest publicly traded energy company. At the back of the hall I wondered how the agency and the peer could be so adamant, when so much of the evidence suggested they were completely wrong.

To start with, the total amount of oil discovered annually has been falling steeply for many decades (figure 9). The best single year for discovery was 1948 when 150 billion barrels were found, most of it in just one field. The exact amount discovered from year to year is of course erratic in the short term, but even when the statistics are smoothed to compensate for this, discovery peaked in 1965. Since then it has been downhill more or less all the way. For almost as long as I have been alive, oil discovery has been in a state of slow-motion collapse.

As a simple run of bad luck this would take some beating, but of course it is nothing of the sort. Contrary to popular conception, there is very little random about the process of oil exploration, and in fact

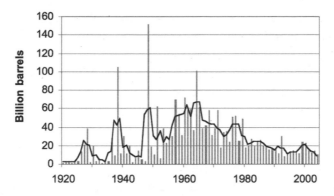

FIGURE 9. The trend in worldwide oil discovery is clear. The biggest single year for discovery (bars) was 1948. When the short-term fluctuations are smoothed out to give a more representative trend (line) the discovery peak comes in 1965.[5] Source: IHS Energy

Hubbert even predicted the worldwide discovery peak in his 1956 paper. He noted, almost in passing, that because the growth in world-wide discovery was beginning to flatten off, 'the period of declining rates of discovery has almost arrived'. Within eight years, it had. And forty years later, by the turn of the twenty-first century, the amount of oil being discovered had dropped to barely one-sixth of its smoothed peak level.

This is not to say that it isn't worth looking any more, since skill, persistence and a good idea can still pay off handsomely. In 1992 petroleum geologist Graham Dore spotted what he thought could be a great prospect in the Moray Firth basin of the North Sea. From studying faint old seismic data, he was convinced he could see a subtle geological trap in deep Jurassic rock that others had missed, and which might be oil-bearing. But his company thought the risk of failure was too high, and refused to investigate further. It took Dore another ten years and a change of employer to get the first exploratory well drilled. Then one Friday afternoon in the spring of 2001 the first tantalizing results came in, although it was another excruciating week before the core samples and well sensor results finally confirmed his hunch. His field had come in, big-time. 'I was completely euphoric, it's the best thing you can do as a geologist. It still gives me goose bumps just thinking about it.'[6]

The champagne was more than justified, since the new field, Buzzard, turned out to be the biggest discovery in the North Sea in the last twenty-five years, with producible reserves of almost 500 million barrels. This sounds like a lot of oil until you put it in the context of world consumption – over 81 million barrels per day in 2004. If the field could be produced as fast as the world guzzles – and of course it couldn't – Buzzard's reserves would last less than a week. In 2004 the world needed to discover the equivalent of sixty Buzzards to meet consumption, but in fact found the equivalent of only twenty.

The fact that Buzzard is seen as a large discovery today is a sign of the extent to which expectations have fallen. The real giants of the North Sea, Brent and Forties, were four to five times the size of Buzzard, but then they were discovered back in the glory days of the 1970s. Even these were fairly modest by historical standards. Most of the world's 'supergiants' – they only get the T-shirt if they originally contained a minimum of 5 billion barrels of producible oil, ten times the size of Buzzard – were discovered long before. Ghawar in Saudi Arabia, which is by far the world's biggest oilfield with 140 billion barrels originally – more than twice as much as will ever be produced from the entire North Sea – was discovered in 1948.[7] The three next largest – Burgan in Kuwait, Safaniya in Saudi, and Samotlor in Russia – were discovered in 1938, 1951 and 1961 respectively. And we are now extraordinarily dependent on a tiny collection of these truly enormous but ageing fields. There are about 45,000 producing oil and gas fields in the world, yet over half the total oil reserves (53 per cent) are held in just 100 fields.[8] All but two of them were discovered before 1970, quite often by geologists using nothing more sophisticated than a surveyor's wooden contraption called a plane table, and a stiff pair of walking boots. Guess where most of them live: the Middle East.

So the sharply declining discovery trend is largely driven by the fact that there are fewer supergiants left to be found. The size of the average-sized discovery confirms the story. Because discoveries are by their nature erratic, it is often better to take a view over several years' exploration effort. In 1940 the average size of newly discovered fields over the previous five years was 1.5 billion barrels; in 1960 it was 300 million barrels; by 2004 it was just 45 million barrels and still falling.[9] This is all exactly as you would expect. Oil explorers are rational, highly skilled and commercially driven. The biggest oilfields are the easiest to find and the most profitable. It's no surprise therefore that the oil industry seeks and discovers them first. There are still lots of smaller fields being discovered, but the total amount of oil they contain cannot match the vast volumes of a few dozen monsters found decades ago.

To find out if there is any chance of these well-established trends being reversed, I consulted Richard Hardman, the former head of exploration and production of Amerada Hess, who during his career was involved in the discovery of 1.2 billion barrels of oil. We met in the grand library in London's Geological Society, of which he was once president, and where they filmed a key scene for the classic 1960s spy thriller *The Ipcress File*. On a massive table between the gloomy rows of leather-bound archives, he unrolled maps and seismic surveys to make his point: 'The world has been surveyed to the extent that it's very unlikely that very large reserves exist in areas that we haven't looked.'[10]

Paradoxically, he explained, it is because the chemistry and geology of oil formation are now so well understood that the situation is so stark. Oil is only found in the sedimentary basins of ancient seas and lakes because these are the only places that offered the right conditions for the remains of micro-organisms to accumulate and be preserved. A number of amazing geological coincidences would then have to occur for the resulting prehistoric sludge to be cooked into oil and trapped in exploitable reservoirs, but without the sedimentary basin you wouldn't have the source material to start with. So there's no point looking anywhere else.

The world's sedimentary basins are obvious geological features, and most of the more prospective have been thoroughly explored. Even the less promising areas have been sufficiently well studied that oilmen can now make a pretty good estimate of what they might hold. 'We have explored virtually all the sedimentary basins that are likely to contain oil, and really for the first time we have a calculation of what reserves they can contain,' says Hardman, 'and it's that calculation that worries me.'

In contrast to the falling rate of discovery, oil consumption has soared since the middle of the twentieth century. From just 10 million barrels per day in 1950, it doubled in the next ten years, and almost tripled again by 1980. The first and second oil shocks caused demand to slump briefly in the recession of the early 1980s, and then to grow at a slower pace as the higher price forced companies and people to use fuel more efficiently. This seemed to knock world oil consumption off what had been its undeniably Hubbertian trajectory – at least for a time. But since the turn of the century the astonishing rise in oil demand caused by rampant economic growth in China and India, combined with continuing over-consumption in America and Europe, has steepened the curve once more.

Despite strong growth in demand, for most of the twentieth century the world discovered far more oil each year than it consumed. But in the mid-1980s the situation reversed and since then we have been pumping much more than we find. For twenty years we having been living off the fat accumulated in the years of plenty, now long gone. These days for every barrel we discover, we now consume at least three.

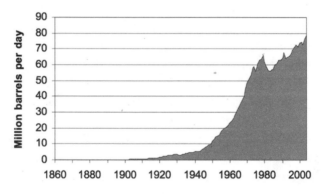

FIGURE 10. Annual world oil production since 1860. Does this shape look familiar? Source: IHS Energy

Since we are consuming so much more than we discover, the obvious next question is how long can this can be sustained, which brings us to the most contentious area of oil industry statistics: reserves. To listen to Lord Browne at the 2005 Oil and Money conference in London, you would think there was nothing to worry about: 'The world holds some 1.1 trillion barrels of oil as well as some 6,200 trillion cubic feet of natural gas which has been found but not yet developed. At current consumption rates that's forty years of oil supply and seventy years of gas.' This sounds like a reassuringly large amount of oil that will last for a long time, but it is a formulation that can be highly misleading.

The idea that the world has x years of oil is based on a simple calculation called the reserves-to-production, or R/P ratio. As the name suggests, you simply take the most recent reserve figure and divide it by the latest annual consumption rate, to come up with the number of years those reserves will last. At first glance this may sound like a reasonable rule of thumb, but it is a formulation that drives Colin Campbell to fury.

Dr Campbell is the forecaster who has arguably done more than any other to wrestle the issue of oil depletion into the public consciousness. After a long and successful international career as a geologist and senior manager with Texaco, BP, Amoco and Fina, he became increasingly convinced that the world was approaching its inevitable oil peak. In the early 1990s he began to model oil depletion using industry data and Hubbertian techniques in collaboration with Jean Laherrère, a former senior exploration geologist for Total. In 1998 they wrote a seminal article in *Scientific American* entitled 'The End of Cheap Oil' – the price has risen almost eightfold since then – and in 2001 they founded the Association for the Study of Peak Oil (ASPO), which today organizes heavily attended international conferences and has branches springing up all over the world.

At his home in the tiny village of Ballydehob in West Cork, Campbell is now used to receiving a stream of foreign visitors including, on one occasion, a group of curious agents from the US Office of Naval Intelligence, complete with standard-issue dark glasses. He explains that while the R/P ratio was a useful rule of thumb for individual oil companies to gauge the security of their reserves in the early days of the industry, it is no longer appropriate. 'In the old days, if an oil company's R/P fell from 10 years to 6, for example, they'd say, "Gosh, we'd better crank up exploration and find some more oil." And in those days the world was big enough and cranking up exploration did deliver. You can't do that any more.' [11] The ratio has lost its rationale, he argues, now that annual discovery represents barely a third of yearly consumption.

Worse still, Campbell maintains, the R/P ratio implies a production profile that is physically impossible. 'Lord Browne gives the impression that production can be held at its current level for forty years and will then collapse to zero in a single year, but this is flat absurd. The peak will come long before that, and oil production will go into irreversible decline for the simple reason that all oilfields decline gradually.'

To get to the bottom of BP's position, I arranged a briefing at the company's London headquarters in January 2005 with Professor Peter Davies, its bullish chief economist. 'You've chosen the wrong subject,' he told me right away, 'we're not going to run out of oil.' It quickly became clear that, as an economist, he views with disdain not only the techniques of peak forecasting, but also the geologists and engineers who use them. When I explain that I would like our discussion to stay on the record he quips, 'I'd better not be too rude about the loonies then.' In defence of Lord Browne, Mr Davies explained, 'He's not saying we're going to run out of oil in forty years, he's saying this is indicative of an adequacy of resources, and that one indicator is a reserves to production ratio of over forty years . . . It's just a number to put 1.1 trillion barrels of oil into context. It seems to be a meaningful indicator.'

In one critical sense, however, it is meaningless. The fundamental problem with the R/P ratio is that it gives no inkling of the approaching peak. In fact because reserves and production tend to decline in tandem, the ratio is likely to stay roughly flat, or even to rise, long after the peak has passed. According to figures from the *BP Statistical Review* in 1995 the UK North Sea had an R/P ratio of 4.3 years; in 1999, the year of peak production, the ratio had risen to 4.7; and by 2004 had risen again to 6 years, despite the fact that production had now fallen by 30 per cent. The same is true for many other oil-producing regions.[12] Yet by the logic of Lord Browne and Peter Davies we should put more faith in a small rise in the R/P ratio than in a collapse in the amount of oil actually being produced. This is evident nonsense.

Even the normally upbeat IEA has private doubts about BP's persistent use of the R/P ratio. The Agency's official view is that there is no peak on the horizon provided that the necessary investments are made in oil production. But at a quiet lunch in London a senior official told me that in his view, 'BP is completely missing the point.' The world R/P ratio of 40 is meaningless, he argued, since Western capital is prevented from investing where most of the oil is – in the Middle East. 'If there was one world it would be OK, but there are barriers. If the money can't meet the oil we are in a very bad situation, and currently it is not happening.' As a result, he said, in his opinion, 'BP is a problem; their story is so rosy that at the end of the day it will hurt those who think this way.'

The R/P ratio may be misleading, but the reserves figure Lord Browne bases it on – 1.1 trillion barrels – is certainly large. Not only that, it is a number that has *grown* over the last two decades (figure 11), which seems to make no sense at all: since production has exceeded discovery since the mid-1980s, reserves ought to be falling, not growing. Yet after two decades in which the world guzzled more than half a trillion barrels, we still seem to have more left over than we started with in 1980. This apparent paradox only resolves itself when you realize that the word 'reserves' embraces several different categories, and that those that sound most reliable are sometimes the most misleading.

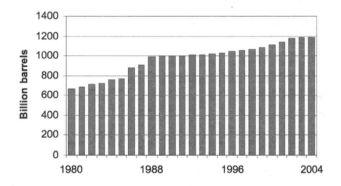

FIGURE 11. World proved oil reserves. Source: *BP Statistical Review*

The two most significant categories are *proved* reserves, and *proved and probable*. Both represent amounts of oil that are known to exist underground, and both are likely to be produced at some point. What distinguishes the two is the degree of confidence the oil company has about this outcome.

This distinction is anything but academic, as Shell discovered in January 2004, when the company was forced to admit that it had previously overstated its reserves by 4 billion barrels. This revelation slashed the value of Shell's shares by £3 billion in a single day, and cost the company its chairman and its reputation. The scandal was not that Shell had invented oil that simply did not exist, but rather that in its desperation to impress the stock market the company had declared as proved reserves oil that ought to have been described as proved and probable. And the great unremarked irony of the affair was that although the stock market fixates on proved reserves, it is proved and probable that generally give a far better guide to the amount of oil that is ultimately likely to come out of the ground.

The reserves with the greatest certainty attached are proved, and these are the figures most commonly cited throughout the industry and the stock market. While the exact definition may vary between jurisdictions, the basic idea is that the oil described as proved is reasonably certain to be produced at current prices and with existing technology. Sometimes the definition is qualified with a probability of 90 per cent, meaning that the company is 90 per cent certain that at least the amount of oil it claims as proved will actually be produced, or that 9 times out of 10 more oil will be produced than the company's estimate. The strictness of the definition is meant to guard against stock market fraud, and means the proved number will almost always be much lower than the amount of oil the company really expects to get out of any given field. But this is the number usually reported in the annual accounts.

Probable reserves are also likely to be produced at current prices and with existing technology, though not as certain as proved. Taken together, proved and probable (2P) are sometimes given a probability of 50 per cent, meaning that the company is 50 per cent certain that at least as much oil as it defines as proved and probable will be produced. Although the certainty is obviously lower than for proved, paradoxically the 2P number is much more likely to be accurate, at least in aggregate. Because the probability is 50 per cent, the company is equally likely to get it wrong by estimating either too high or too low, so over a large population of fields the errors should cancel out. Proved and probable are effectively the company's best guess of what its fields will in fact produce, but most big companies tend not to publish these numbers in their annual accounts.

To grasp the real significance of these distinctions, it helps to understand how the estimates are produced. I consulted Trevor Ridley, a wry South African-born petroleum engineer who trained at Imperial

College and then spent much of his career assessing reserves for Amoco in Trinidad, Texas and Norway. His new company Corsair Petroleum is raising funds to develop various North Sea fields, which is how I came to meet him at a 'prospect fair' in London.

Whenever an exploration company strikes oil, it is the petroleum engineer's job to assess how much there is. The company's geologists will already have identified and mapped a promising underground structure, otherwise they wouldn't be drilling there in the first place, and now there are sensor measurements and core samples from the first well. These not only confirm the presence of oil, but also allow technicians to gauge the porosity of the reservoir rock, and the proportion of the pores occupied by oil and gas. Now the engineer can multiply these numbers by the volume of the whole reservoir to find the total amount of oil down there – the *original oil in place* (OOIP).

The OOIP can be described as the 'resource' but never as 'reserves', because in the average field most of the oil will never leave the reservoir. But furnished with the information from the first well the company now knows something about the geology of the field, and from past experience can estimate the *recovery factor* – what proportion of the oil they will eventually get out. This in turn helps the company to estimate both proved and probable, and proved.

To calculate the proved and probable reserves, the engineer may simply multiply the original oil in place by the expected recovery factor. If the OOIP is 100 million barrels and the recovery factor 35 per cent, then proved and probable is 35 million barrels. At first this is only a rough estimate and will change as the engineer's understanding of the geology improves, but it is his best guess at how much oil will eventually be produced from the whole field. 'If you do this properly,' Ridley explains, 'in the long run it's proved and probable that should be closest to what eventually comes out.'

When the engineer comes to calculate proved reserves, however, he has to be much more cautious, because he must be reasonably certain that the figure he produces will be equalled or even surpassed in practice. This forces him to limit the area of the field he claims as proved (figure 12). 'Basically you draw a dotted line around your well, saying "I am confident that the conditions we have observed in the well will exist out to the limit I have just drawn."' And since there is only one well to start with, the proved reserves figure can be as little as a tenth of the proved and probable figure in the early days of a field's development. What this means is that as more wells are drilled, proved reserves will almost inevitably rise. 'It's exactly what you'd expect,' says Ridley, 'but it doesn't mean you are finding any more oil.' As the field is progressively drilled up, proved reserves will rise towards the proved and probable ceiling, giving a false impression of growth.

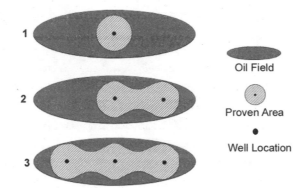

FIGURE 12. Aerial view of a hypothetical oilfield showing the evolution of proved reserves as more wells are drilled. When the field is first discovered, its proven area is typically a small fraction of the total area of the field, and its unproven area contains 'probable' reserves. The second and third wells extend the proven area of the field, but the field itself is of course still the same size, and the total of 'proved and probable' is unchanged

This impression is perpetuated in the global statistics. The proved reserve numbers in the *BP Statistical Review*, so frequently cited by Lord Browne, are world proved reserves. They are 'true' as far as they go, but figure 13 puts them into context: although proved reserves have risen since 1980, proved and probable figures from IHS Energy, a consultancy that maintains one of the industry's leading databases, have been falling for almost as long.[13] This properly reflects the fact that we have been consuming much more than we have discovered. The graph lines of the two sets of reserve figures have now converged so much that they are practically kissing, and this might suggest that growth in worldwide proved reserves is almost over, as indeed it may be. Notice in particular how much more slowly proved reserves have grown in recent years.

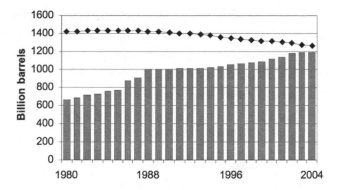

FIGURE 13. Global proved reserves (bars) give the reassuring appearance of continuing growth, but the more relevant proved and probable (diamonds) have been falling since the mid-1980s. Sources: *BP Statistical Review;* IHS Energy

In fact the graph lines probably should not be quite so close, because they are based on different assumptions and sets of data. Whereas the IHS figures are constructed by gathering data on thousands of individual fields from a variety of industry and official sources, the *BP Statistical Review* simply regurgitates the reserve numbers claimed by governments, many of which are inflated for political purposes. For instance, the sharp upward step in the BP proved numbers visible in the mid-1980s was the result of Middle East OPEC countries raising their reserves as part of a dispute over quota allocations, and most observers believe the increase is largely bogus. So from the early 1990s onwards the proved reserves ought to be lower than shown.

For a petroleum engineer there is no doubt about which are the most telling numbers, and that their message is clear. 'The proved numbers are absolutely misleading,' says Trevor Ridley, 'but the IHS numbers have all sorts of meaning attached to them, because as proved and probable they are the best estimate of the remaining oil available for production. These figures tell you that we are running out of oil, and the slope tells you how quickly we're running out.'

Roger Bentley, an academic who founded the Oil Resource Group at Reading University, has studied the intricacies of reserve reporting for the best part of a decade, and warns that *all* oil industry data is unreliable, but nevertheless agrees: 'The proved reserves tell us absolutely nothing, and it's their spurious growth over the years that has lulled everybody into a false sense of security. You get a much better sense from the IHS proved and probable numbers, and they clearly show the critical turning point back in the early eighties when we started to consume more than we discovered. They make it quite obvious that supply difficulties must be very close.' [14]

Just to make things complicated, however, there is another kind of reserve growth, what you might call *real* reserves growth, although the extent and significance of this growth is hotly disputed. According to the United States Geological Survey, BP and others, proved and probable reserves also grow over time. And unlike proved reserves, where 'growth' is simply the result of the distortions of oil company reporting practice, the proponents of this kind of reserves growth argue that it results in a genuine increase in the amount of oil available to humanity. 'The number of physical barrels in the world really is irrelevant,' declares BP's chief economist Peter Davies. 'The amount of *recoverable* oil is a function of economics and technology, which are not fixed. If the price rises from $40 to $80 the amount of oil you can produce is very different.'

According to this view, advancing technology constantly enables the industry to exploit new patches of oil that were previously inaccessible – by drilling in ever deeper water off West Africa for instance – so increasing the size of the total resource. Technological progress is also claimed to enable companies to extract an ever larger proportion of the resource by increasing the recovery factor. And should a shortage ever really bite, the theory goes, the oil price would rise and allow the exploitation of previously uneconomic fields, so solving the problem.

These views are widely held, especially among economists, and superficially seem to be supported by the facts. The oil industry has made some stunning technological advances, and it does manage to squeeze more oil from the rock as time goes on. According to the IHS Energy database, the amount of oil added to known resources over the last decade by these factors has been greater than the amount added by discovery.[15] But unfortunately the idea that technology or price will prevent oil production from hitting a peak then going into terminal decline is just not supported by the history.

Jim Henry's whole career has borne witness to that history. Every inch the Texas oilman, from his beautifully tooled boots to the cream-coloured Stetson on the back seat of his 4x4, Jim was studying petroleum engineering at Oklahoma University when Hubbert uttered his famous prediction in 1956. Fifteen years later, just as Hubbert's prediction was coming true, Jim left the oil company he was working for and founded his own. Henry Petroleum is now one of the region's most successful independents, with 1,200 wells on production. For Jim and his generation of oilmen life has been a constant battle between depletion and technology.

Jim drives me out of Midland, the oil capital of West Texas, to see some technology in action, the shooting of a three-dimensional seismic survey.[16] In the middle of a field a line of futuristic-looking trucks lower vibrating pads on to the ground to send shockwaves deep into the earth's crust, to be reflected by the geological strata and picked up by sensors miles away. The technique builds up a minutely detailed picture of the geology below, revealing three tiny reservoirs, each about the size of a football field, which it turns out are too small to be economic. Three-dimensional seismic may allow oilmen to find a needle in a haystack, geologically speaking, but it can't change the fact that there are only needles to be found. 'A lot of people think all this new technology is going to save us, but it doesn't work that way,' says Jim.

Of all the oil-producing regions in the world, America has enjoyed the best access both to capital and the most advanced technology. In his long career, Jim has seen it all: water flooding, CO_2 injection, 3-D seismic, 4-D seismic, horizontal drilling, multilateral wells, reservoir fracturing, visualization suites, and ultra-deepwater drilling in the Gulf of Mexico. But despite all the technological advances, throughout Jim's career oil production in the United States has declined, year after relentless year. The reason, according to the former president of the Permian Basin Petroleum Association, is simple: 'In the natural course of events we find the huge reservoirs first. In the US we found lots of the biggest in the thirties and forties. And when they start declining, because their production rates are so huge, we can't make up the difference with all the little fields we're finding today,' Jim explains. 'Technology can kind of mitigate the decline, keep the decline from being so steep, but it won't stop the decline.'

It's a similar story in the North Sea. In 2000 I took a trip to the Saltire production platform, 120 miles north-east of Aberdeen, owned by Talisman, a Canadian independent that specializes in mature oilfields. Talisman's UK managing director Paul Blakeley explained that they had installed new equipment that would allow them to drill three or four times further than before, and 'that makes a big difference to tapping into new reserves of oil.'[17]

That's not three or four times *deeper*, but sideways. The Extended Reach Drilling equipment allowed the drill bit to turn at a sharp angle under the sea bed and burrow not only thousands of feet down, but also outwards from the rig to a distance of perhaps five miles. It would then drill horizontally to create in effect several different wells within the thin geological layer they were trying to tap. In his cabin the drilling director, surrounded by high-tech screens, resembles a NASA controller.

This kind of drilling allows Talisman – and the rest of the industry – to exploit many of the remaining small reservoirs in the area efficiently from a single platform, which would be uneconomic if each required its own rig. 'It's technology and innovation that gives the North Sea a bright future,' Mr Blakeley declared during my visit in 2000, just as the province passed its all-time peak. But by 2005 North Sea oil production had already fallen 30 per cent, at which rate output from the largest province discovered since the Second World War will drop to less than a fifth of its peak level by 2020, with the vaunted technology apparently powerless to stop it.

In the years since my visit Talisman had kept production from Saltire steady by drilling several more wells, and by buying fields from other operators the company had managed to expand its output within a declining province. But although Mr Blakeley remains optimistic about future North Sea output, even he now accepts the terms of reference have changed: 'Success in the North Sea isn't about growing production, it's about slowing down the decline.'

If technology is not quite what it's cracked up to be, neither does the economists' cherished price mechanism seem to have as much impact as they would have you believe. In 2002 Harry Longwell, then the director and Executive Vice-President in charge of exploration and production for ExxonMobil, wrote that 'contrary to some widely held beliefs, discovered volumes, over a long period of time have not been closely related to price fluctuations . . . In the recent past, we have seen increasing demand for oil and gas, but generally decreasing discovery volumes, during a period of fluctuating but generally higher average prices.'[18] This is something of an understatement: the price of oil has risen from less than \$2 per barrel in the 1960s to almost \$80 more recently, while discovery has plummeted over the same period.

Neither does price appear to have a fundamental bearing on oil production once a province has passed its geologically imposed peak. Figure 14 compares the trajectories of British North Sea output since its peak in 1999 and the oil price. The only correlation that immediately comes to mind is inverse: declining North Sea production contributes to a worsening world supply-demand balance, so helping to drive the price upwards. In any event, the spot price rose eightfold between 1999 and 2006, yet production plunged by more than 30 per cent.

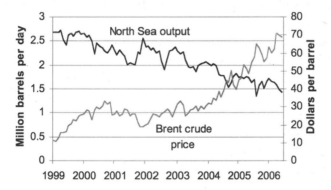

FIGURE 14. North Sea oil production continues to slide despite the soaring price of Brent crude. The oil price shown in this graph is lower than the record daily highs reported in the news over the same period because it is based on a monthly average. Sources: Royal Bank of Scotland; Argus Media

Since oil industry investments have long lead times it may be that the severity of the fall partly reflects the delayed impact of the previous collapse in the oil price in the late 1990s. But there is no doubt that the decline is fundamentally geologically driven, and that changes in technology and economics make a relatively marginal difference. Even BP's Peter Davies seems to accept that whatever happens North Sea production will continue to fall: 'The geology is well picked over,' he admits. 'We can affect the gradient but we can't affect the direction.' Coming from an arch-opponent of the peak oil camp this seems to be a significant concession; if what he says is true for the North Sea, it will apply everywhere.

Despite this, Mr Davies sees no sign of an imminent global peak. On the contrary, he argues repeatedly that since peak forecasts have been made in the past and proved wrong, the basic methodology must be faulty. 'If you're going to listen to these people you have to look at their track record, and if they've been consistently wrong you have to ask yourself why. And is there something new out there which says this is now different?' It soon becomes clear that for Davies this question is purely rhetorical. 'It's all been said before, the doomsters are wrong and the economists can simply declare victory.' And, he concludes with a dismissively Blairite flourish, 'Let's move on.'

It is perfectly true that several forecasts of the global peak have come and gone without the sky falling in. In the 1970s Hubbert predicted the global peak would arrive in 2000, and more recently Colin Campbell has made a number of forecasts that have likewise turned out to be premature. It is also true that the weakness of Hubbert's simplest method – which Campbell still employs – is that it depends on an estimate of ultimate, which in turn is vulnerable to the assumptions of the modeller and the quality of the data.[19] Campbell acknowledges this candidly: 'The one thing you can say about my numbers is that they're wrong; perhaps Mr Davies would like to tell me by how much.' What is interesting today however is that while many peak forecasts use or imply resource estimates that are far higher than Campbell's current figure, they still produce broadly similar results. This even includes the biggest and most optimistic resource assessment of the lot.

In 2000 the United States Geological Survey released its *World Petroleum Assessment* to howls of protest from peak forecasters. The USGS had concluded that between 1996 and 2025 oil discovery could total 649 billion barrels, and that 'real' reserve growth would deliver almost as much again – 612 billion – implying a conventional ultimate of just over 3 trillion barrels.[20] Many analysts were incredulous at the numbers and suspected that the USGS, with its history of over-optimistic resource assessments, was up to its old tricks again. But it turns out that the Survey has very helpfully calibrated the entire debate in a way that, perhaps inadvertently, supports the case for an early peak.

In 2005 the USGS published a second report assessing how well the forecasts contained in the earlier work had performed against reality, and found that their reserve growth forecast was on track but that discovery was badly adrift.[21] To satisfy the Survey's discovery total of 649 billion barrels, the world would need to find almost 22 billion barrels every year until 2025, but during the first quarter of the forecast period (1996–2003) we discovered just 9 billion a year on average. And since we know that oil discovery is in long-term decline, this deficit is only likely to worsen in future. This has the effect of knocking off perhaps 400 billion barrels or more from the USGS figure of 3 trillion for ultimate. A conventional ultimate of 2,600 billion barrels is in the same ballpark as the figures used by most of the forecasters who predict a global peak within a tight range of about fifteen years. What's more, we have already consumed close to half of this figure. And since it includes a generous allowance for reserve growth, there is nowhere else for the economists to hide. Let's move on.

Another reason why peak predictions are likely to be more accurate today is that the areas of uncertainty are shrinking fast, a point made to me forcefully by Jason Nunn, a British director of the Washington-based consultancy PFC Energy. 'In the seventies and eighties when everybody said we were going to run out of oil, very few countries were in plateau, let alone decline. Today the majority of countries are in plateau, and many are in decline. Thirty years ago many parts of the world had not been explored, whereas today there are very few countries that haven't been. It's different now.' [22]

What is most strikingly different is that the pattern witnessed by Jim Henry in America is now asserting itself all around the world. And it is this more than anything that convinces me of the imminence of the global peak. Whatever you make of the obscure technical arguments about how best to analyse the reserves, depletion is no longer a distant problem; it is demonstrably happening in the here and now. The production numbers are undeniable, and the list of countries in terminal decline is growing almost by the year.

Each year the *BP Statistical Review* lists the output of about fifty of the most significant oil-producing countries, and according to an authoritative analysis of the figures by the *Petroleum Review*, eighteen have now passed their peak and are falling at an average of 3.5 per cent a year.[23] Some never amounted to much, but many are prolific producers such as Venezuela, Norway, Indonesia, Oman, Argentina, Egypt, Australia and Colombia, as well as the United States and Britain (figure 15). Collectively, production from this grouping has fallen from 22.8 million barrels per day in 1996 to 19 million barrels per day in 2005. According to the *Petroleum Review* several other world-class players have either just peaked or will do so soon, including Mexico, Denmark, Turkmenistan and Brunei. And the complete list of post-peak countries, including all the small ones, numbers more than sixty.[24] Despite the technological advances and despite the rising oil price these green bottles just keep on falling.

Until now these declines have been more than offset by production growth in other parts of the world. Of the fifty-odd significant producers listed in the *BP Statistical Review*, over thirty were still growing in 2005, and there was particularly strong growth from countries such as Azerbaijan, Thailand and several West African states. But these strong individual performances look rather different when seen in the context of a regional analysis.

If you divide the world into its three main oil-producing blocs (OPEC, the countries of the former Soviet Union, and the rest of the world), it becomes clear that 'the rest' has been struggling for years (figure 16). Despite the strong growth in many individual countries, total output of the non-OPEC, non-former Soviet Union bloc has stayed just about flat since 1998. For all the billions of dollars invested by the oil industry in that period, all the new production barely matched the fall in output from the declining countries. In fact, for the OECD (the industrialized countries, and a subset of 'the rest'), total output has been falling since 1997.

As world oil demand took off at the turn of the century, for a few years it was effectively satisfied only by booming production in Russia – where output growth has now slowed to a crawl – and OPEC. But by 2005 even they were out of spare production capacity; Katrina and Rita simply let the outside world in on the secret.

Part of the problem that the hurricanes exposed and worsened is undoubtedly above ground: refinery capacity is limited; a worldwide dearth of drilling rigs is holding back exploration programmes; and the industry is critically short of petroleum geologists and engineers. All this is serious enough, because none of these factors will improve quickly: refinery investment is now picking up but major increases in capacity take several years to build; each offshore drilling rig

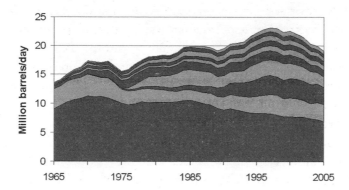

FIGURE 15. Oil production in the ten largest post-peak countries, in size order: United States, Venezuela, Norway, UK, Indonesia, Oman, Argentina, Egypt, Australia and Colombia. Collectively they produced 22.8 mb/d in 1996, and 19.8 mb/d in 2005. Source: *BP Statistical Review*

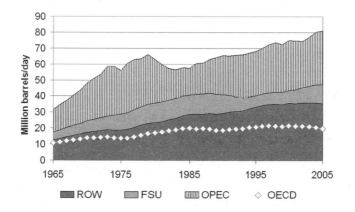

FIGURE 16. Global oil production by bloc. Output from the non-OPEC, non-FSU world (ROW) has been flat since 1998, and all growth has effectively come from Russia and OPEC. OECD production has been in decline since 1997. Source: *BP Statistical Review*

costs hundreds of millions of dollars and also takes several years to construct; the workforce is greying, and the industry has real difficulty recruiting into what is widely seen as a dirty relic of the 'old economy'. All these factors weaken oil companies' ability to fight depletion.

But the fundamental problem is clearly underground. ExxonMobil, Total and oil services company Schlumberger all estimate that the average decline rate in the world's oilfields is 5 per cent, meaning that the industry has to build 4 million barrels a day of new production capacity every year simply in order to stand still.[25] That's the equivalent of adding more than twice Britain's 2005 oil production capacity every twelve months, before the industry can even start to satisfy any demand growth. At the same time discoveries continue to fall, new countries pass their production peak almost every year, and by most estimates of the world's ultimately recoverable oil we are rapidly approaching the halfway mark. All of which suggests it cannot be long before the geology finally imposes the wrong kind of shortage – the sort you can do very little about.

4

Long Fuse, Short Fuse

IT MUST HAVE seemed like a good idea at the time. On 16 February 2005, the day the Kyoto Treaty came into force, thirty Greenpeace activists invaded the trading floor of the International Petroleum Exchange in London with the aim of bringing the $6.5 billion-a-day business to a halt.[1] The protesters' plan was to make it impossible for the 'open outcry' pit traders to bawl their deals to one another by setting off rape alarms attached to helium balloons. What they hadn't counted on was the level of exception the dealers would take to this interruption of their money making after a long liquid lunch.

'They just went wild,' said one activist, 'they were trying so hard to hit us they were falling over each other.'[2] In the mêlée the protesters found themselves pummelled into a corner, where Greenpeace director Stephen Tindale was almost relieved to have a metal bookcase pulled down on his head: 'That wasn't great,' he told me, 'but it made a kind of barrier between us and them.' Eventually the protesters were forced out on to the pavement, and one was taken to hospital with a suspected broken jaw. Floor trading was interrupted for just over an hour, although electronic trading continued throughout.[3] So what was the point?

The protesters had also scaled the front of the building and hoisted a banner that neatly summarized their thinking: 'Climate change kills. Stop pushing oil'. The second sentence is the key; in their view global warming is largely the oil industry's fault. Yes, we all consume oil products, but for Greenpeace the oil companies are the big villains because they continue to invest tens of billions of dollars in oil production against relatively tiny amounts in renewable energy. So they believe that making progress beyond the first modest step of Kyoto will involve forcing the *industry* to mend its ways and *sell us less oil*.

That afternoon a Greenpeace spokesman told me excitedly, 'We need huge cuts in CO_2 emissions. We had to tell oil companies that additions to oil have got to stop!' In other words, far from worrying about oil running out, these people are willing it to.

And not without reason. There is now absolutely no doubt that global warming is a desperately serious problem. The science is good and the impacts are already obvious: the record hot summers just keep on coming; 35,000 people died in a single European heatwave in 2003; tropical storms are becoming ever more violent; the ice caps and glaciers are melting, and the sea level rising at twice the pre-industrial rate; and the frequency with which the Thames Barrier has had to be raised to prevent flooding has jumped from once every two years in the 1980s to six times a year more recently.[4] Looking ahead the worsening effects are expected to include droughts, falling crop yields, hunger, the geographical spread of tropical diseases, and the displacement of tens of millions of people as low-lying coastal areas such as in Bangladesh are flooded. Many of the world's major cities are threatened.

As if all that wasn't bad enough, the outlook for the coming century could be truly catastrophic. The Intergovernmental Panel on Climate Change (IPCC) foresees average temperatures rising between 1.4°C and 5.8°C by 2100.[5] To put this in context, many scientists now believe that a bout of global warming within that range caused the Permian mass extinction 250 million years ago. This was the mother of all mass extinctions, and far more devastating than the one that wiped out the dinosaurs 65 million years ago. A BBC *Horizon* documentary on the subject was aptly entitled *The Day the World Nearly Died*. Initially, massive volcanic eruptions in Siberia emitted enough CO_2 to raise temperatures by about 5°C, which would have been deadly for many species, but could not explain the full extent of the mass extinction. What it did, however, was to warm the oceans sufficiently to release billions of tonnes of frozen methane from the sea bed, and this led to a further 5°C increase in temperatures, which very nearly finished the job. It made the seas stagnate, turned the atmosphere toxic and wiped out 95 per cent of all species on the planet.

Global warming today is already triggering the same kind of 'positive feedback loops' that drove the Permian mass extinction, although their impact was not included in the IPCC models which produced the 1.4–5.8°C forecasts. In August 2005 scientists reported that a vast area of Western Siberia's frozen peat bog – the size of France and Germany combined – was beginning to thaw for the first time in 11,000 years.[6] Like the sea bed, this so-called permafrost contains billions of tonnes of methane which could now be released. According to Professor Paul Wignall of Leeds University, the scientist who pieced together much of the jigsaw of Permian mass extinction, 'Once you start to release methane hydrates in any quantity, the world is on a spiral to destruction.'[7] We are clearly already travelling in fantastically dangerous territory.

So it might seem that the last oil shock is a blessing in disguise. After the global peak, some forecasters expect oil production to fall by about 3 per cent per year, and by coincidence this is almost precisely the figure by which emissions need to fall in order to stabilize the atmospheric concentration of CO_2 at 450 parts per million and restrict warming to about 2°C this century. In 2005 Friends of the Earth launched a campaign – www.thebigask.com – calling for the introduction of a law obliging the government to enforce reductions of 3 per cent a year. But could nature be about to render the campaign obsolete, by suppressing one of the main causes of global warming in the nick of time?

The question is of course rhetorical, because oil accounts for less than 40 per cent of total CO_2 emissions, with the rest coming from coal and gas. The International Energy Agency (IEA), which maintains there is no global oil peak on the horizon, predicts that unless government policies change, consumption of oil, gas and coal will continue to rise at least until 2030, and that as a result CO_2 emissions will soar by 55 per cent. However, my analysis of the IEA's figures shows that if oil production were to peak – say – in 2010 and then decline at 3 per cent per year, oil consumption and its associated emissions would fall by almost 40 per cent by 2030, but total emissions would still rise by almost 24 per cent.[8] This is not counting the even greater quantities of coal and gas that would presumably be burnt in a desperate attempt to make up for the dwindling supplies of oil. So cutting our oil consumption – whether we do so voluntarily or frogmarched by the last oil shock – is absolutely necessary to fight climate change, but nothing like sufficient. Oil may take care of itself, but coal and gas emissions must also be cut hard and these reductions will probably have to be self-imposed.

You might expect that climate change campaigners would embrace the idea of the last oil shock. After all, it gives an even greater urgency to do many of the things they are already lobbying for, such as developing alternatives to oil. But two words you will never hear most of them utter in public are 'peak oil', even though they are by now well aware of the debate. During a lengthy interview with Greenpeace director Stephen Tindale in March 2005 it became clear that the reasons for this reticence are a mix of political calculation and a certain – how shall I put this? – confusion about the rudiments of oil depletion.

Tindale accepted that the argument might provide extra ammunition in support of policy proposals that would also be good for global warming: 'The empirical evidence is that politicians do not respond to climate imperatives; perhaps they might respond to peak oil imperatives.' Yet he was convinced it would be a mistake for his organization to campaign on the issue. 'This is a highly contested area and there is nothing approaching a consensus,' he argued. 'If it turned out that the oil peak thesis was wrong, and we'd been using it, then that would undermine and discredit other things we had been saying.'

For a group of people who have struggled for years to drag climate change up the public policy agenda, this caution is understandable, but their avowed distrust of the science of oil depletion is by now insupportable, especially when it seems to be based on depletion illiteracy. There is no excuse for Stephen Tindale to be defending such a position on the basis of statements like 'A major new oil find would presumably change the facts,' and 'I think as the oil price goes up more oil will be found.' As we saw in the last chapter these views, although widely held, are profoundly wrong. This level of ignorance among groups whose overwhelming concern is energy is hard to understand.

Although Greenpeace professes scepticism about the science of the last oil shock, they would love the event to happen, and soon. Tindale summed up: 'Frankly, let's hope that the oil does run out and that oil prices go up and that the world has to develop alternatives to oil seriously quickly, and from a climate point of view that would be an excellent outcome.'

You can only hold this view with equanimity if you believe there are alternatives to oil ready to deploy right away. Remember, the last oil shock is not principally about power generation – replacing coal-fired power stations with windmills and so on – but overwhelmingly about transport fuels; oil supplies 95 per cent of all transport energy.[9] In order to embrace the arrival of the last oil shock this warmly you must be confident of developing alternative fuels to replace a growing proportion of the 80-plus million barrels of crude oil we currently consume every day. But climate change imposes another constraint – the replacement fuels must also emit no CO_2, or at least very much less – and this severely limits the options. Strictly speaking there are only two alternatives that could satisfy the criteria: hydrogen produced from non-CO_2 emitting sources and biofuels. The key questions now are whether these resources are big enough and can be developed quickly enough to mitigate either climate change or the last oil shock.

Arnold Schwarzenegger is quite the most improbable environmentalist. As a movie star his career mostly consisted of blowing things up, and as a politician he has displayed all the subtlety you would expect, memorably denouncing his opponents as 'girly men'. He is also credited with persuading General Motors to develop the Hummer, that grotesque, overtly militaristic SUV based on the US forces' Humvee troop carrier. Dealers report that the H2 model did less than 10mpg.[10] Arnie has a collection of such vehicles, but interestingly one of his H2s actually runs on H_2: hydrogen.

The truck doesn't actually belong to him, being a prototype on loan from General Motors, who developed the hydrogen Hummer at undisclosed cost, and who insist on one of their engineers riding shotgun whenever it is driven. But this is not simply a rich man's indulgence; surprisingly it is also a token of one of Governor Schwarzenegger's key policies. In 2004 he launched the California Hydrogen Highway Network, a $100 million programme to subsidize the opening of up to 200 hydrogen fuelling stations by 2010, giving every resident ready access to what he clearly believes is the solution to our various energy crises. 'Californians invent the future and we are about to do it again,' he declared. 'We have the opportunity to prove to the world that a thriving environment and economy can co-exist.'

Six months later the Governor rolled up to a hydrogen filling station at Los Angeles International Airport in his Hummer and refuelled for the cameras. *Forbes* magazine described the addition of a hydrogen engine to such a four-tonne monster as putting 'lipstick on the pig', but it was also an undeniably potent symbol: it said we can continue to squander energy in absurdly unnecessary ways and *still* save the planet.[11]

The man behind Arnie's Hydrogen Highway is his Cabinet Secretary Terry Tamminen, a former environmental charity advocate whom Schwarzenegger appointed to run his administration. We talk in Tamminen's cramped office in the Sacramento Capitol, with the bicycle he rides three miles to work each day propped against the wall.

It soon became clear that one of the main motives for the Hydrogen Highway is an acute awareness of looming oil shortages. Tamminen told me in September 2005 that California was now a net importer of petroleum products, the world's refinery capacity is at full stretch, and China and India are now 'gobbling up more and more of the scarce resource'. (The words pot, kettle and black sprang to mind at this point, but I let it pass.) As a result, Tamminen predicts petrol shortages and a $10 gallon by 2010 – 'if you can get it!' He is a 'big believer in Hubbert's peak', but argued that short-term, non-geological constraints such as terrorism or political upheaval in producing countries may overtake the significance of the idea: 'It may not matter what's left in the ground if we can't get at it'. And the *only* solution to all this, argues Tamminen, is hydrogen.

Cars that run on hydrogen already exist; I took a test drive at the California Fuel Cell Partnership, the Sacramento research centre where a collection of car makers evaluate the performance of their various prototypes. From the outside you would never have known my car was anything but an ordinary Ford Focus – apart from the words 'Hydrogen fuel cell electric' emblazoned on the rear bumper. The ride was also just like a normal automatic: it accelerated smoothly, and they tell me the top speed is about 85mph. The only noticeable difference was the noise, or rather lack of it. The electric motor gave off nothing but a gentle whine, rather like an airport shuttle train, and the fuel cell itself is entirely silent.

The principle of the hydrogen fuel cell was discovered by a Swiss scientist in 1838, but it was only in the 1960s that the device found its first really practical application, providing electricity and water in spacecraft. Here's how it works. Hydrogen is the most abundant element in the universe but on earth is never found on its own. It has a particular attraction to oxygen, and when combined they form water. The fuel cell works by exploiting this electrochemical bond.

Oxygen and hydrogen are fed into two chambers in the fuel cell, separated only by a membrane coated with a catalyst such as platinum. The catalyst causes the hydrogen to split into its constituent protons and electrons, and the membrane only allows the protons to pass through to join the oxygen. The electrons are also desperate to join the party, but since the doorman won't let them in, they're forced to sneak round the back. The alley they duck into is a circuit which bypasses the membrane, so generating an electrical current, and which then reunites them with the hydrogen protons and the oxygen. The current drives an electric motor that powers the car, there is no combustion of carbon and therefore no CO_2 emissions, and the only waste product is water.

There is no doubt that the idea is both simple and elegant. Better still, the overall efficiency of a hydrogen fuel cell is more than twice that of a conventional internal combustion engine. In an ordinary car, only about 18 per cent of the energy contained in the fuel actually reaches the tarmac to do anything useful, the rest being wasted through inevitable friction and heat loss in the engine and transmission. In a fuel cell car – with fewer moving parts – the figure can be as high as 50 per cent. If this were the only consideration, as enthusiasts sometimes imply, there would be very little argument.

One serious obstacle at the moment is cost. During the test drive I turn to my host Mike Smith and ask what it would cost to buy one of Ford's fuel cell cars today. He grinned from under his wraparound shades, and replied, 'Oh, about a million dollars.' But he wasn't joking.

In part the car is so expensive simply because it is a prototype, and costs would obviously fall with mass production. But many of the components – not least the precious metals needed for the fuel cell catalysts – are inherently expensive. And there is real caution among senior management in the industry about when or even whether the costs of fuel cell vehicles will match those of traditional cars. Yozo Kami, the engineer in charge of Honda's fuel cell programme, has admitted it will take at least a decade to bring the sticker price of their model down to $100,000.[12]

The next issue is safety. Hydrogen has had a lousy reputation on this front ever since the *Hindenburg* went up in 1937, although some argue that the gas was not to blame for the tragedy. According to a controversial analysis by former NASA scientist Addison Bain, it wasn't the hydrogen that ignited – at least not at first – but the lacquer that had been used to dope the skin of the airship. It turns out they had painted it with what would soon be used as the principal components of rocket fuel. *Doh!* Naturally, hydrogen enthusiasts today rather like this theory.

Nevertheless hydrogen deserves its reputation. Since the molecule is so tiny, hydrogen is by definition more likely to leak than other gases such as methane, and it requires much less energy to ignite. One study found that in the event of a leak of pressurized hydrogen, a spark of static electricity caused by the flow of gas could cause ignition.[13] When Mike refuelled the Ford, he not only connected the hydrogen nozzle, and a lead to transmit data between the station and the car, *but also an earth cable*. I was asked to stand twenty feet back to take a photo of the process.

The next big problem lies in the boot of the car. In fact, as Mike opens it up, I see the problem occupies almost the entire space: it's the hydrogen fuel tank. Since hydrogen gas is so diffuse it must be pressurized to at least 5,000 pounds per square inch to hold anything like enough energy within a reasonable volume. And because of the pressure the tank must be cylindrical and extremely heavy. This one is about three feet long by two and a half feet wide, and would take you about 200 miles. But as things stand, don't expect to pack much luggage. And I'm not sure how eager most people will be to put their children in a car packing hydrogen at 5,000psi, no matter how good the crash test results.[14]

Any one of these knotty problems might well be enough to stymie the emergence of a hydrogen-based transport system for decades, although enthusiasts like Terry Tamminen insist they will all be overcome: 'There are no show-stoppers here.' But even if the problems of cost, hydrogen storage, safety and customer acceptance were all solved tomorrow, there would still remain some far more fundamental drawbacks.

Hydrogen is not actually a fuel but a vector, or energy *carrier*, rather like electricity. It may be the most common element in the universe, but you can't just dig it up somewhere and burn it; it has to be prised apart from its molecular companions and then transformed in various ways. Virtually all the hydrogen produced today (96 per cent) is made from oil, coal or natural gas, a process in which the hydrocarbon provides both the feedstock and the necessary heat and power. This is the cheapest way to produce hydrogen, but the process consumes lots of energy and generates lots of greenhouse gas emissions.

Almost half the hydrogen produced today comes from natural gas, where the process is 70–90 per cent efficient, depending on the size of the plant. Let's call it 80 per cent on average, meaning that 20 per cent of the energy content of the methane is consumed in the process.[15] But hydrogen gas at a remote petrochemical plant is not much use to anyone. The best way to get it to a filling station, if you don't want to build an expensive pipeline network, is by road tanker. That means liquefying the gas, which involves cooling it to at least −240°C, just a few degrees above absolute zero. This process is about 70 per cent efficient, meaning it consumes the equivalent of 30 per cent of the remaining energy.

Like Ford, most car manufacturers have concluded that the best way to store hydrogen on board is not in super-cooled liquid form, but as compressed gas. So at filling stations such as the California Fuel Cell Partnership in Sacramento, the liquid hydrogen has to be allowed to revaporize, before being compressed again. The process of compressing hydrogen to 5000psi is about 90 per cent efficient, meaning it consumes the equivalent of about 10 per cent of the remaining energy.[16] Since the fuel cell vehicle itself is at best 50 per cent efficient, it loses half the remaining energy. In total, a staggering 75 per cent is wasted, and scarcely a quarter does any useful work by moving the wheels over the tarmac:

$$100 \times 0.8 \times 0.7 \times 0.9 \times 0.5 = 25.2\%$$

As a result of all this, fuel cell vehicles at the moment are *less* efficient on a well-to-wheels basis than petrol hybrids. Toyota calculates that the Prius is 32 per cent efficient overall, while its compressed hydrogen fuel cell vehicle is just 29 per cent efficient. And so much CO_2 is emitted by the hydrogen production and transport process that, according to a study by Malcolm Weiss of the Massachusetts Institute of Technology, there is very little to choose between fuel cell vehicles and their toughest competitors running on petrol.[17] The report concluded that although fuel cell vehicles would emit much less in the way of greenhouse gases than normal cars, petrol hybrids like the Prius already match those reductions, and the diesel hybrids expected to be launched in the next couple of years would beat them. So why bother?

For the supporters of fuel cell vehicles like Terry Tamminen, hydrogen from natural gas is simply a stepping stone, a way of subsidizing the spread of the technology until truly clean methods of producing hydrogen can be developed. The ultimate goal is to make hydrogen by electrolysing water, using non-CO_2 emitting sources such as wind, wave or solar power. In theory this could produce limitless quantities of greenhouse-gas-and-guilt-free transport fuels, and even wean the world off hydrocarbons altogether. But although the sun and the wind seem to be more or less infinite sources of energy, even here there are profound limitations.

To start with, making hydrogen by electrolysis will always be far more expensive than producing it from natural gas, since it takes about ten times as much energy to break the bonds in a molecule of water as it does in one of methane. That's why only 4 per cent of hydrogen is made in this way. (To give you an idea of how much energy electrolysis demands, two chlorine plants in Cheshire that operate by electrolysing salt water consume as much electricity as the entire city of Liverpool. Or at least they did until 2005, when the soaring gas price forced them to cut back production.)[18]

Producing hydrogen by electrolysis consumes the equivalent of 35 per cent of the energy contained in the end product. Even if it was possible to make the hydrogen locally at the filling station, so doing away with the need to liquefy the gas for distribution, the compression would still consume another 10 per cent and the fuel cell another 50 per cent. In total, fuel cell vehicles run on hydrogen produced locally by electrolysis would still waste more than two-thirds of the original energy:

$$100 \times 0.65 \times 0.9 \times 0.5 = 29.25\%$$

But why should we care? Such wastage would of course make the new fuel expensive, but on the other hand we would no longer have to worry about oil and gas running short, the process would emit little or no CO_2, and supplies would be infinite. At least, that's what the supporters of hydrogen would have you believe. What they tend to ignore, however, is that the inefficiency of electrolysis, combined with the fact that renewable sources of electricity are so diffuse, means that the energy infrastructure needed to supply such a transport system would have to be truly immense.

Let's say you wanted to run the entire British road vehicle fleet on hydrogen fuel cells, and to be able to drive the same distances as before. Assume the fuel cell vehicles will use hydrogen electrolysed from water at local filling stations, using non-CO_2 emitting sources of electricity. At the moment there isn't a single fuel cell lorry in existence, so there are no mileage or efficiency figures to help us to work out what would be required to power the crucial haulage sector – at least not directly. But we do have the numbers from the prototypes of fuel cell cars, and these provide a good starting point for the calculation.

In 2004 British cars and taxis travelled a total of 400 billion miles (table 1). Ford claims the fuel cell Focus that I took for a test drive does 52 miles per kilogram of hydrogen, and Honda claims that its 2005 FCX fuel cell car does 57, so lets call it 55mpkg on average. That means British motorists would need almost 7.3 billion kilograms to do the same mileage they did in 2004.

Each kilogram requires 65 kilowatt hours (kWh) of electricity to produce – enough to power a one-bar fire for almost three days – giving a total of nearly 473 billion kWh (40kWh is the energy contained in the kilo of hydrogen, the rest is what's used in electrolysis and compression). Divide total kilowatt hours by the number of hours in the year, and it turns out that Britain would need additional electricity generating capacity of 54 gigawatts. But we haven't finished yet.

Private motoring consumes two-thirds of Britain's road fuel, and freight accounts for the rest. Although there aren't any hydrogen fuel cell lorries, let's imagine there were. Let's also make the reasonable assumption that the ratio of fuel consumption between the private motoring and the freight sectors stays roughly the same; trucks will

Table 1: How much wind, solar, nuclear to replace current US and UK road fuel via hydrogen?

DEMAND	US	UK
Mileage (2004)	2,570,400,000,000	400,000,000,000[a]
Divided by mileage per kilogram of hydrogen	55	55[b]
Gives kilograms of hydrogen required	46,734,545,455	7,272,727,273
Multiplied by kWh electricity needed to electrolyse and compress 1 kg hydrogen	65	65[c]
Gives kWh required	3,037,745,454,545	472,727,272,727
Expressed as MWh	3,037,745,455	472,727,273
Divided by the number of hours in the year (8,760) to give MW capacity required	346,775	53,964
Raised by 50% to account for freight gives total generating capacity required	520,162	80,946[d]

WIND	US	UK
Generating capacity needed	520,162	80,946
Divided by individual turbine capacity (3MW) gives theoretical number of turbines	173,387	26,982
Multiplied by a factor to compensate for average output of just 30% rated capacity	3.33	3.33
Gives actual number of turbines needed	577,380	89,851
Divided by the maximum number of turbines per km² (4), gives area required in km²	144,345	22,463[e]
Expressed as square miles	55,732	8,673

SOLAR	US	UK
Generating capacity needed	520,162	80,946
Divided by capacity (MW) of 1km² solar array	100	100[f]
Gives theoretical area needed in km²	5,202	809
Multiplied by factor to compensate for average productivity (14% US, 8.6% UK)	7.00	11.90[g]
Gives actual area needed in km²	36,411	9,633
Expressed as square miles	14,058	3,719

NUCLEAR	US	UK
Generating capacity needed (MW)	520,162	80,946
Divided by Sizewell B capacity (MW)	1,200	1,200
Gives number of Sizewell Bs required	433	67
Or as multiple of total current nuclear capacity	5	7[h]

Notes on sources and conversion factors can be found in footnote.[19]

obviously continue to consume more fuel per mile since they carry heavy loads.

In order to convert our figure of 54 gigawatts, which is equivalent to 66 per cent of current fuel consumption, to one that equates to total fuel consumption including freight, we have to raise it by about half the original figure, giving 81 gigawatts. That's more than Britain's total current generating capacity.[20] Where would we get this much juice?

The common-or-garden wind turbine, the kind that measures about 400 feet tall and 260 feet across the blades and gets little Englanders all hot and bothered, has a peak capacity rating of 2.5 megawatts (MW). But higher capacity turbines are now being introduced, so let's be generous and assume 3MW as standard. Divide 81 gigawatts by 3MW and in theory Britain would need 27,000 turbines. But since the wind blows at different speeds, and sometimes not at all, on average these windmills produce no more than 30 per cent of their rated capacity. So that number has to be multiplied by about 3.3, giving a grand total of almost 90,000 turbines. Since you can fit no more than four of these windmills in a square kilometre, this implies the wind farm would cover almost 22,500 square kilometres (8,700 square miles), which is a good bit bigger than Wales, or even the entire south-west of England. I personally find wind turbines rather elegant, but imagine having one every 500 metres in every direction across Devon, Cornwall, Dorset, Gloucestershire, Somerset and Wiltshire. The rest of the sky, of course, would be thick with pylons and high voltage cables.

To produce the same electricity from solar panels would cover a land area roughly equal to North Yorkshire or, if you prefer, Norfolk and Derbyshire combined. (That's every inch of it, by the way, not just the rooftops, so you wouldn't be growing any turnips.) Alternatively, you could opt for sixty-seven additional nuclear power stations with the capacity of Sizewell B (at the moment we have the equivalent of just ten).

America obviously has more space, but here too the exercise throws up some fairly extraordinary numbers. Their wind farm would have to be bigger than the land area of Illinois, Wisconsin, New York state or Iowa. It would even take four-fifths of a large, empty state like North Dakota. Alternatively they could build a solar farm that would more than cover New Jersey, Massachusetts or Maryland, or knock up 430 Sizewells.

When I put the American examples to Terry Tamminen he laughed and quipped, 'I think we could give up North Dakota in return for energy security, don't you?' And no doubt the Jeremy Clarkson riposte to the British example would be similar, although probably cruder: 'So what? Sod the Welsh, get building.' But this is to miss the point entirely.

Let's leave aside the question of whether or not we actually have enough space in Britain for this lunatic scheme, ignore the fact that it would mean doubling the capacity of the national electricity grid, fondly imagine that there was an infinite supply of engineers, and kid ourselves that not one objection would be lodged at a single planning enquiry. Even if none of the above was a problem, how long do you think it would take to achieve?

The British Wind Energy Association hopes the industry will expand capacity by about 800MW per year for the rest of this decade. But even if they managed to double that rate, it would take fifty years to reach the target. Total installed solar capacity in Britain is just 10MW, which is 0.01 per cent of what would be required. All but three of Britain's ageing nuclear power stations are due to close by 2014, and despite the government's decision in the 2006 Energy Review to back nuclear new-build, a nuclear-hydrogen revolution looks impossible. Because planning and building a reactor takes so long, the industry is unlikely even to replace the current reactors as they close, let alone raise total nuclear capacity *sevenfold* any time soon. Meanwhile, once oil production peaks, our daily supply could almost halve in twenty years.

Joe Romm is as big a detractor of hydrogen as Tamminen is an advocate. It's what you might expect from the author of a book called *The Hype About Hydrogen*, but in fact he was once an enthusiast too. As a senior official with the US Department of Energy in the Clinton administration Romm oversaw a massive increase in funding for hydrogen research, and at first he intended his book to be a primer not a dissection. But the more he investigated the more sceptical he became, and by the time I interviewed him at home in Washington in September 2005 he was adamant: 'Hydrogen is the very last alternative fuel you would try if all others fail.'

For him the fundamental objection is not simply that fuel cell cars won't be marketable for at least a decade, nor just the huge waste of primary energy, nor even the massive and expensive new infrastructure that would be required, but crucially that renewable sources of energy could be employed in ways that would fight global warming far more effectively.

Any hydrogen produced in the way I have sketched out would displace oil, but if the windmills and solar panels were used to displace coal and gas being burned in power stations, it would reduce CO_2 emissions much more. So using renewables to make hydrogen for transport fuels would actually make matters relatively *worse*, even though the hydrogen itself was being produced cleanly. According to a study conducted by the Energy Saving Trust, using renewable electricity to replace gas-fired power would save 100g of carbon per kilowatt hour (C/kWh), whereas using it to replace road fuels would save only 60gC/kWh.[21] The authors concluded: 'Until there is a surplus of renewable electricity it is not beneficial in terms of carbon reduction to use renewable electricity to produce hydrogen – for use in vehicles, or elsewhere.'

In these circumstances, Romm argues with some passion that to produce hydrogen for fuel cell vehicles on any scale would be 'criminal', and a perverse strategy for dealing with our approaching 'energy train wreck'. Hydrogen as a transport fuel is by now so *obviously* the wrong answer that Romm questions the US government's motives in backing it. The $1.2 billion that President Bush committed to hydrogen research in his 2003 State of the Union speech 'so that the first car driven by a child born today could be powered by hydrogen, and pollution-free' was a temporizing manoeuvre by an administration that has no real intention of cutting greenhouse gas emissions or even US oil import dependency.[22] 'It is a smokescreen, placebo, bait and switch, pick your favourite metaphor. This isn't to say there aren't people like Terry Tamminen who don't sincerely believe in hydrogen, and I'm not questioning their motives, but I am questioning the motives of the administration and General Motors. They don't want fuel efficiency standards, and are holding up an imaginary techno-fix. It is 95 per cent cynical.'

In short, hydrogen as a transport fuel seems to be utterly incapable of mitigating either global warming or the last oil shock. No doubt it will make a few headlines and find some useful niche roles – in Iceland for instance, where they have limitless hydro-electricity and a population of about 300,000 if you don't count the haddock – but for the rest of the world, it's back to the drawing board.

At first Jones the Grocer noticed an unusually brisk trade in his cut-price cooking oil. Then Davies the Plod fancied he could detect a whiff of chip fat among the traffic fumes. The Revenue got suspicious and set up a taskforce with the local police – inevitably dubbed the Frying Squad. It was the great Llanelli fuel scam of 2002.

It all started with the petrol protests in 2000 when, starved of fuel at the pumps, resourceful locals in South Wales began to experiment with blends of diesel and vegetable oil which they knocked up in the garage at home. To their delight some found that their diesel cars ran just as well if not better than before, and many kept on using the newly discovered biofuel even after the protests ended. At forty-two pence a litre you could see why. But since the oil was going into a fuel tank rather than a chip pan they were meant to be paying duty. Enthusiasm for the idea waned after the Frying Squad started impounding cars and fining their owners £500.

But it wasn't the end of cooking oil as fuel. Asda, the supermarket that first noticed the jump in sales in its branches in South Wales, now recycles all the spent frying oil from its canteens to produce a biodiesel blend to run its fleet of delivery vans. And in Norfolk, a Thetford-based company called Global Commodities now has the capacity to produce 350 million litres of diesel a year, much of it from used cooking oil.

The company was founded by local entrepreneur Dennis Thouless. He was about to retire but one of his grandchildren was 'doing the environment' at school and badgering him to 'put something back'. So he started to look into the idea of using second-hand cooking oil for domestic heating, but after provoking his wife's displeasure by blowing up their central heating boiler in the depths of winter, he turned his attention to transport. Here the research paid off and in 2001 he sold all his business interests to sink into the new venture.

The company buys up used cooking oil, mostly from big food-processing companies such as crisp manufacturers, and refines it into biodiesel which is then sold to local garages and hauliers. The fuel is extremely green, says Thouless, because very little extra energy is required to turn what would otherwise be waste into perfectly good diesel. The only snag is that the total amount of used cooking oil available in Britain each year is estimated to be about 300 million litres, whereas annual diesel sales are more than 25 billion litres.[23] The sad fact is that we just don't eat enough fish and chips to save the planet this way.

Old chip fat is not the only way to make biofuel. But while there are many different potential sources, there are just two basic varieties. *Biodiesel* can be made from more or less any oil seed crop such as sunflower, rapeseed or soybean, which are processed into fuels known as methyl esters. These can either be blended with fossil diesel or used neat in diesel engines without modification. *Bioethanol*, on the other hand, is made by fermenting sugar from crops such as wheat, maize, sugar beet or cane. This can be used in petrol engines either blended or neat, although existing vehicles have to be modified to run on higher proportions of ethanol. At the moment most car manufacturers' warranties are only valid up to a 10 per cent blend. The other main distinction is that while biodiesel contains the same amount of energy by volume as traditional diesel, a litre of ethanol contains only two-thirds the energy of a litre of petrol.

Both are 'green' fuels in the sense that they are produced from plants which themselves absorb CO_2 as they grow, but the extent to which they actually reduce emissions depends critically on the production methods employed. If the crop has a relatively low yield, or its production demands heavy use of fertilizers or machinery run on fossil fuels, the benefits are reduced. Assessing quite how much the various biofuels could cut greenhouse gas emissions is a complicated and controversial business.

However, the biggest problem by far with biofuels is the amount of space required to produce them, and this brings energy demand into direct conflict with food production. The European Union has set a non-binding target that 5.75 per cent of road fuels should come from renewable sources by 2010, and the UK government has set its own target of 5 per cent by the same date. Yet a study by the International Energy Agency has found that replacing just 5 per cent of European petrol and diesel consumption with biofuels would consume the output of 20 per cent of its cropland.[24] So even hitting these modest, initial targets will have an impact on food production, which could only worsen if biofuel production were to rise further.

The IEA figures also imply that even if we devoted *all* our cropland to biofuel production we would only produce a quarter of our current fuel consumption. We could all starve to death in a traffic jam. They also make it quite clear that if biofuels are to play a really significant role in future, they will certainly have to be grown in other, sunnier, more spacious parts of the world. So the questions demanded by the last oil shock are where, how much, and how quickly.

Even for North America the sums don't seem to stack up. The United States is already a big producer of bioethanol, although it's hard to see why they bother, except perhaps as a sop to the powerful farming lobby. Even with $80 oil the business has to be subsidized, and the process is so inefficient that the resulting fuel contains scarcely more energy than it takes to produce.[25] The resource constraints are just as tight in the US as in Europe. A study by researchers at the University of Minnesota found that if the entire US maize crop was converted to ethanol it would supply only 12 per cent of the country's petrol consumption.[26] Nevertheless, American ethanol production continues to soar and is estimated to have consumed a fifth of the US maize harvest in 2006. The furious pace of expansion of US ethanol production has caused grain prices to spiral, and could soon start to eat into the level of grain exports from the world's largest producer.

And it's not as if the world has any food production to spare. The veteran environmental campaigner Lester Brown reported in 2006 that the world grain harvest has failed to match consumption in six out of the last seven years, and that stocks are dwindling to dangerously low levels. In this context the increasing diversion of maize for ethanol production will create 'an epic competition between the 800 million motorists who want to protect their mobility and the two billion poorest people in the world who simply want to survive'.[27] The grain required to produce enough ethanol to fill a twenty-five-gallon SUV fuel tank would feed one person for a year, yet the amount being diverted into ethanol production continues to soar, reflected in the spiralling price of maize.

The conflict between biofuels and food production may be eased, however, by a new method of producing ethanol that is just in its infancy, known as cellulosic ethanol. A Canadian firm called Iogen has developed a small demonstration plant in Ottawa and is planning to scale up its operations in Canada, America and Europe, with the new industrial-scale plants due to start production in 2009. The technology is evidently promising, since a sizeable stake in the company has been bought up by Shell, an act of 'great foresight' according to the oil giant's former chairman Lord Oxburgh.

Ron Ox – as he signs off his emails – was never going to be your typical oil company boss. His formidable curriculum vitae is largely academic – he taught geology and geophysics at both Oxford and Cambridge – and includes a five-year stint as chief scientific adviser to the Ministry of Defence, and a spell as rector of Imperial College London. After being parachuted in to Shell to restore its reputation after the reserves scandal, he infuriated his press handlers with his blunt opinions on climate change, telling journalists that unless CO_2 emissions were dealt with, there was 'little hope for the world'.

Over tea in the Peers' Guest Room at the House of Lords in November 2005, Oxburgh is still full of surprises, confessing that as an undergraduate he used to down pints with Colin Campbell, now the grand old man of peak oil. 'In a fundamental sense he's quite right,' says Oxburgh, although he adds – as I pick myself up off the floor – 'but I think his interpretation of his logical position is probably too extreme.' Notice the 'probably' in that sentence. Perhaps the ever courteous peer is just being customarily polite, or perhaps he's not so certain.

Lord Oxburgh argues that what we are likely to see is not a peak, but a plateau. As crude oil becomes 'almost too expensive to burn', alternative fuels will be developed in a relatively few years to keep overall liquid fuel supplies at least flat – a horizontal line on the graph. I suggest to him that even if he's right about overall liquid fuel production – and I don't believe he is – the word 'plateau' is deceptively reassuring. It conjures up a broad expanse of land that is easy to travel across safely, whereas a far better analogy for such a horizontal graph line would be 'tightrope'. He laughs and takes the point, but insists that biofuels in particular can be brought on stream in fairly short order, and will be a 'major assistance in smoothing out the humps in the plateau, and perhaps even filling some of the chasms under the tightrope'. He is particularly enthusiastic about the Iogen cellulosic ethanol process, because its feedstock is agricultural 'waste material'.

Unlike traditional ethanol production the new process consumes not the edible part of the crop, but the fibrous parts that are 'left over', such as cereal straw or corn stover. Genetically modified enzymes break down the cellulose in the straw first into sugars and then into ethanol. The process is rather neat because once the cellulose has been broken down the remains of the feedstock can be burned to provide most of the heat and power needed for the production process. As a result Iogen claims its cellulosic ethanol generates 'virtually no' net CO_2 emissions. Lord Oxburgh agrees that the fuel's environmental footprint will be 'almost immeasurably small' and that it will cost as little as about $20 per barrel to produce: 'It is probably the best biofuel in commercial and environmental terms that anyone has ever come up with.'

The problem, again, is the size of the resource. Straw is neither free nor strictly a waste product. A large proportion has to be ploughed back into the fields to maintain soil structure and help with water retention, and yet more is needed for animal bedding and other farm uses. As a result only a fraction of the straw actually produced would normally be available. Iogen claims that each tonne of feedstock produces roughly 300 litres of ethanol, and one study by an institute set up to promote biofuels estimates that 17.8 million tonnes of crop residues are available in Canada each year, giving a production potential of 5.34 billion litres. Since ethanol contains just two-thirds of the energy of petrol, this equates to 3.5 billion litres of gasoline, less than a tenth of Canada's annual consumption.[28]

Even the enthusiastic Lord Oxburgh concedes that the process is largely a way of exploiting a particular situation in big cereal-producing countries such as Canada and the US. Although he foresees a time when all waste containing carbon-hydrogen bonds – such as timber, household rubbish and sewage – will also be used to produce fuel in this way, it is not a silver bullet solution. 'This of itself is not going to solve all of the fuel problems of the future. You are not talking about the complete replacement of existing needs.'

Ethanol production makes much more sense in the tropics, where the sun's rays are stronger, and the energy yields from sugar cane far better. Brazil first started to promote ethanol as a fuel after the first oil shock, when the price of oil soared and the price of sugar, its principal crop, collapsed. Today all fuel sold in Brazil contains between 20 and 25 per cent ethanol, and 100 per cent ethanol fuel is widely available. Overall ethanol provided 40 per cent of Brazil's non-diesel road fuel in 2004, and Flex-Fuel cars, which can run on either petrol or ethanol or any combination of the two, made up more than half of new car sales in 2005.[29] You simply fill up the tank with whatever combination of fuels you choose, and a clever piece of kit in the exhaust pipe discerns what mix the car is running on and recalibrates the spark timing and fuel injection automatically.

Brazil has the cheapest and most energy-efficient ethanol production in the world. The fuel delivers eight times more energy than the fossil fuels that go into producing it (against 1.3–1.8 at best for US corn-produced ethanol), and in late 2005 it was selling for the equivalent of 55 US cents a litre without subsidies.[30] This is partly because the sugar cane yields are good, and partly because the bagasse residue from the cane is burned in combined heat and power plants to provide the energy needed to refine the sugar. In fact Brazilian sugar processors generate about 600MW (half a Sizewell B) of surplus power which is sold into its national grid. For these reasons the CO_2 savings are also substantial.

Brazil has well-advanced plans to become a big exporter. Its agriculture minister declared in 2005, 'We don't want to sell litres of ethanol, we want to sell rivers,' and the government claims that there are an astonishing 90 million hectares (900,000 square kilometres) available for sugar cane production in the interior of the country. By my calculation (table 2) this could *in theory* produce the equivalent of just under 6 million barrels of petrol per day, equivalent to 29 per cent of world consumption in 2003.

There are several problems with this however. First, the land that the Brazilian government is talking about is part of the country's vast Cerrado uplands, a wild savannah that was originally twice the size of Western Europe, and an area of great biodiversity and ecological importance. The 90 million hectares is what's left, and converting all

Table 2: Brazilian bioethanol potential

Litres bioethanol produced (2004)	15,000,000,000[a]
Divided by land area employed, hectares	2,660,000[b]
Gives average litres per hectare	5,639
Total land available, hectares	90,000,000[c]
Multiplied by current yield of	5,639
Gives total potential litres	522,518,796,992
Multiplied by factor to reflect lower ethanol energy content	0.66[d]
Gives gasoline equivalent litres	343,559,822,331
Expressed as barrels/day	5,919,873
World total motor gasoline demand (2003), barrels per day	20,248,770[e]
Brazil ethanol potential as percentage of 2003 world gasoline demand	29
Hectares required to supply 2003 world motor gasoline demand	319,677,000
Annual planting required to match 3% oil decline rate, hectares	9,590,310

Notes on sources and conversion factors can be found in footnote.[31]

of it to sugar cane production would not only threaten thousands of native plant and animal species, but also the sources of some of Brazil's biggest rivers. According to Professor Jim Ratter of the Royal Botanic Garden Edinburgh who has studied the Cerrado for forty years, 'The effects on climate and hydrology could be very, very severe.'[32]

The fact that it could be an ecological disaster does not of course mean that it won't happen. But the scale of the undertaking does mean that it is unlikely to happen quickly.[33] Ninety million hectares is well over thirty times the area of Brazilian land currently producing sugar for ethanol, and one and a half as big as Brazil's total cultivated crop-land. To exploit it would mean building a massive infrastructure of roads, warehouses, liquid storage facilities and power lines, which would all take time. Dr Paulo Wrobel, an energy adviser at the commercial section of the Brazilian Embassy in London, says, 'I don't see any possibility of this happening in the next ten to twenty years. I'm speculating here but it could take fifty to a hundred.'[34] He also stresses that the 90 million hectares figure is not a target, but merely intended to illustrate that Brazil has the potential capacity to be a major exporter of ethanol in future.

On the basis of my rough calculation, to replace all of the world's 2003 petrol consumption would take 320 million hectares, more than fifteen times the world total of land cultivated with sugar cane in 2004.[35] The IEA predicts that oil demand will rise by more than a third by 2030, which would increase the land required by another 100 million hectares.[36]

According to Dr Jeremy Woods of Imperial College, whom I interviewed in June 2006, many countries in the tropics and sub-tropics do have large amounts of scrubby woodland that are suitable for sugar cane: 'Large areas of Africa and South America have good growing characteristics, are not heavily populated, and have quite a lot of water – which is the other big issue.' He estimates that 150 million hectares are available altogether, and lists Angola, Mozambique and Tanzania as countries with the greatest potential. But Woods is also adamant that this in itself is not a solution to oil depletion, and not just because of the shortfall in land area. 'There are huge environmental, social and land use issues to be resolved,' he explains, 'and it is critical that any biofuel production is developed sustainably.'

The history so far is not encouraging, because despite the availability of large areas of suitable land, biofuel development has mostly involved the destruction of vast swathes of tropical forest. Malaysia and Indonesia (with Chinese funding) are already doing this to create enormous palm oil plantations, which threaten the survival of the orangutan and the pygmy elephant, as well as the sources of major rivers.[37] Palm oil is an ingredient in a wide variety of processed foods, cosmetics and detergents, but it is also beginning to be used for biofuels.

The academic and environmental commentator George Monbiot has argued persuasively that even if biofuel plantations are hacked out of virgin jungle, the markets they create will inevitably compete with food production elsewhere, and almost certainly increase hunger around the world:

> The market responds to money, not need. People who own cars have more money than people at risk of starvation. In a contest between their demand for fuel and poor people's demand for food, the car-owners win every time. Something very much like this is happening already. Though 800 million people are permanently malnourished, the global increase in crop production is being used to feed animals: the number of livestock on earth has quintupled since 1950. The reason is that those who buy meat and dairy products have more purchasing power than those who buy only subsistence crops.[38]

It is difficult to argue with that logic, but in a debate about the adequacy of fuel supplies, grotesquely, it's almost beside the point. Even if we were prepared to tolerate the ecological destruction and the injustice of global poverty and hunger, it is hard to see how such an enormous transformation could happen fast enough to combat global warming, let alone keep up with the last oil shock. There may be large areas of land available for biofuels production, but to match a 3 per cent oil decline rate would mean planting well over 9 million hectares of sugar cane *every year* – a land area equal to Sri Lanka and Equatorial Guinea combined. What's more, as inadequate as the sugar cane strategy might be, it fails to address less than half of the problem: ethanol replaces petrol, whereas more than half the world's road fuel is supplied by diesel.

In some respects biodiesel could prove far less damaging than ethanol. Diesel can be made from ecologically destructive palm oil plantations, but it can also be made from the nuts of a hardy tree *jatropha curcas*, which can apparently thrive on marginal land anywhere with a tropical or sub-tropical climate. A small British company called D1 Oils has plans to plant millions of hectares of the crop in India, China, Thailand, Ghana and Zambia, and, during an interview in London in November 2005, chief executive Philip Wood explained that the plant's traits mean they will avoid the traditional pitfalls of biofuel agriculture. 'When environmentalists say biodiesel is bad because it's taking land out of food production for the starving millions in the world to feed fat Germans' Mercedes, that may be true if it's made from rape, or soya or what have you, but it isn't with jatropha.'

According to D1 Oils, jatropha can be grown on poor soils, with little water or fertilizer, and so need not compete with food production, nor encroach on rainforests – in fact it positively prefers relatively arid conditions. 'It's not magic,' says Wood, 'but it does seem particularly hardy, long lasting and low maintenance.' The idea is to train poor subsistence farmers in the developing world to produce the nuts on their marginal land, and to refine the oil locally in units so small they can be transported on the back of a truck. Waste water could be used for irrigation, and fertilizer would come from local compost or the seedcake that is left over from the refinery process – which could also be burned to provide electricity. The farmers would produce under contracts approved by their local governments, and the deal 'should provide a significant increase in their income', according to Wood. Depending on the view of individual countries, about half the biodiesel would be kept for the local market, and the rest of it exported.

Although the ecological and social impacts may prove more benign than for other biofuels, the numbers required to replace global diesel consumption are just as daunting. D1's plans are most advanced in India, where they already have several tens of thousands hectares planted. The government claims the country has 40 million hectares of such 'wasteland', on which Wood reckons it could produce about 50 per cent more than current domestic diesel consumption. And other small energy consumers with large areas of marginal land such as Madagascar could easily do the same. But by my calculation (table 3), to replace *global* diesel consumption would need almost 370 million hectares. Even if that much 'wasteland' exists in the tropics, the rate at which it would need to be planted is even greater than for sugar cane. To match a post-peak oil decline of 3 per cent would mean planting almost 11 million hectares, a land area the size of Cuba, every year. D1's target is to plant a total of 5 million hectares globally in ten years.

After we thrash through the figures, Philip Wood sums up: 'The conclusion you come to is that we're not going to solve the world's

Table 3: Jatropha required to replace post-peak oil decline

World diesel demand, 2003, tonnes	971,406,000[a]
Divided by average jatropha yield, tonnes of biodiesel per hectare	2.7[b]
Gives hectares needed to replace 2003 world diesel demand	359,780,000
Annual planting required to match 3% oil decline rate, hectares	10,793,400

Notes on sources and conversion factors can be found in footnote.[39]

energy crisis by planting biodiesel. Fundamentally one has to reduce consumption as well as find alternative sources.'

In the air there are even fewer viable alternatives to fossil fuels than on land. The world uses about 4½ million barrels of jet kerosene each day, equivalent to about one-eighth of daily global road fuel consumption.[40] This would be a daunting enough figure to try to replace, but even leaving aside any resource constraints the technological barriers are also formidable. A study by researchers at Imperial College in 2003 concluded that there are only three potential renewable alternatives to fossil fuels in aviation, all of which would be more expensive, and none of which, according to principal author Bob Saynor, is likely to be available soon.[41]

Jet engines could fairly easily be adapted to run on hydrogen, but airframes would have to be completely redesigned to accommodate far bigger fuel tanks. The most likely designs would be of the 'flying wing' type, but these are a long way off: 'We're talking decades rather than years,' says Saynor. Biofuels – even if you had enough land – are no use to airlines, since they turn viscous at the low temperatures encountered at altitude. 'If you get a blocked injector in a Land Rover you can just pull over, but in the air it's a bit more serious.' And the last option considered by the Imperial researchers, using the Fischer-Tropsch chemical process to turn biomass such as wood chips into synthetic kerosene, would be too expensive and in any case better reserved for road vehicles. Saynor sums up: 'In the short to medium term we don't have realistic and economically viable options.'

And that's it. Hydrogen and biofuels are the only transport fuels that even putatively tackle both global warming and oil depletion, and they appear to be completely inadequate to either task. Every other option – making liquid fuels from gas or coal using Fischer-Tropsch, as the US airforce plans to do, or running electric vehicles from nuclear power stations – will either worsen global warming or emerge too slowly to mitigate the last oil shock. So I return to the question: how can the environmentalists possibly remain so sanguine about depletion?

There are a few who get it, however. Jeremy Leggett, the former Greenpeace chief scientist and author of *The Carbon War*, the classic insider's account of the battles that led to the Kyoto agreement, is now convinced that the last oil shock will be a crisis every bit as profound as climate change: 'They're both problems that threaten the future as we know it, so I wouldn't like to draw a distinction. They certainly have different length fuses, but they're both large weapons of mass destruction.' And when I suggest that since peak oil evidently has the shorter fuse, we should be concentrating on that: 'Yes, of course, absolutely. If we were sensible, logical, joined-up thinkers, which we're not.'

Leggett took his time coming round to the idea, despite being a petroleum geologist himself, and the story of his 'conversion' is instructive. For over a decade he lectured at the Royal School of Mines at Imperial College in London, turning out hundreds of young oil explorers, giving barely a second thought to depletion. But towards the end of the 1980s, as the evidence about global warming was becoming undeniable, he found he could no longer reconcile himself to his job, and jumped at the chance to join Greenpeace. In 1998 Leggett remembers reading Campbell and Lahererre's seminal *Scientific American* article, 'The End of Cheap Oil', but still not believing that a peak was imminent: 'It's a cultural thing; I had worked in the oil industry.' What finally convinced him was the Shell reserves scandal that broke in January 2004. 'I thought then, This doesn't smell good, and resolved to start digging.' He quickly concluded that Shell's troubles were simply a symptom of a far deeper problem, as he argues in his second book, *Half Gone*.

In a café in Waterloo opposite the offices of his company solar-century, during an interview in March 2005, Leggett explained that his former environmentalist colleagues continue to resist the idea of the last oil shock, also for cultural reasons. But in their case it is largely driven by their hostility towards oil companies: 'A lot of environmentalists prefer the opposition to be black; all sides in this debate struggle with shades of grey. That's got a lot to do with it.' So for them the idea that oil might soon start to run short is an awkward complication: 'They want to believe there's lots of this stuff because then you've got a well-defined enemy and the arguments are simpler.'

This resistance to the urgency of oil depletion not only distorts some environmental groups' tactical approach ('Stop Pushing Oil'), but also leads them into fundamental mistakes of analysis and policy – further dissected in chapter 10. Greenpeace director Stephen Tindale, for instance, told me that the ultimate answer to the problem of replacing fossil fuels in transport is hydrogen. He also argued that climate change could be conquered without forsaking our cars and gadgets, and without substantial cuts in our standard of living – with the exception of flying. Neither of these positions looks tenable once you accept that global oil production will soon go into terminal decline.

By contrast Leggett believes the last oil shock will be a 'profound economic dislocation'. And perhaps surprisingly for the chief executive of a company that makes solar panels, he is bleak about the power of renewables to ease the transition. 'If we had more of a running start things might have been different. But I can't conjure up any formulation of alternative energy that can plug the gap in this period. Notwithstanding all the encouraging stuff, I still can't see how you can pump it up quickly enough to make a difference.' When the short fuse of the last oil shock burns down to its charge, it will expose our reliance on oil in profound and unexpected ways.

5

Last Oil Shock, First Principles

I FIND A frivolous but striking image of our utter oil dependency in Galveston Bay, the broad and sheltered inlet on the Texas Gulf coast which funnels tankers into the Houston Ship Channel and the refineries that line its banks all the way up to the city, forty miles inland. Just after sunrise one steamy late September morning I am hammering north in tiny flat-bottomed Boston whaler with three middle-aged men who have taken the day off work to pursue their grand obsession. After about fifteen minutes we reach the buoy that tells them we're where we need to be, the skipper James Fulbright cuts the engine, and we wait.

John, whose weathered face and long blond hair don't quite square with his career in property management, scans the horizon through heavy yellow binoculars. We are a couple of miles out in the bay, but further still from the open sea, and sheltered from the choppy Gulf of Mexico by the sandbar of Galveston. Early morning sun glances off the surface of the murky green water slapping gently against the boat. The two surfboards jammed between the hull and the wheelhouse seem oddly superfluous. Peter, a shaven-headed lieutenant in the local Sheriff's Department, investigates the ice-box. James takes in some rays. I fiddle with my camera.

Finally, John spots what we've all been waiting for: 'Here comes the motherlode,' he drawls in his thick Texan accent, 'snub-nosed and low in the water!' A heavily laden oil tanker has just rounded the point and is steaming fast in our direction. Quickly they slip into a well-rehearsed routine. John takes the wheel and puts the boat – which has drifted – back in position. The other two toss the surfboards over the side and dive in after them. As the 800-foot carrier *Isabella* draws level, the swell generated by her 105,000 deadweight tonnes hits the shallows and rears up to create a wave. Paddling hard to catch it, moments later Peter and James are on their feet and surfing. And unlike the beach – where a surfer's ecstasy is measured in seconds – here they're still riding *two miles*, and *ten minutes* later, as the tanker wave just keeps rolling over the shoals. Their all-time record is 4½ nautical miles in 22½ minutes. 'It's a real leg cramper,' James tells me afterwards. 'You're totally spent physically and mentally. It's insane!'

It took Fulbright, a board shaper and documentary film maker from Galveston, years of patient research to find the exact locations out in the bay where the tanker waves form – weekends spent drinking with gruff old shrimp fishermen, poring over navigational charts, and out in his boat – and I am expected to keep my mouth shut. The crew doesn't only surf the bow waves of supertankers – car carriers, container ships, anything fast and heavy will do – but they describe their passion as 'tanker love' for a reason. 'I hate to say it,' says James, 'but I want more tankers. Bring it on!'

Not only do the tanker surfers owe their ride to the ship's precious cargo – albeit indirectly through displaced seawater – but almost everything else they use to pursue their secretive thrill also comes from crude. The polystyrene foam that gives the boards their buoyancy is made of oil, as are the resins that form their rigid outer shell, and the plastic fins that make them manoeuvrable. So too are the surfers' brightly coloured nylon 'boardies', the baggy shorts favoured by the subculture, and in the winter they'll be grateful for their oil-derived neoprene wetsuits. And let's not forget the drinks cooler, the sunshade over the wheelhouse, sunglasses, sunblock, flip-flops, cameras and lenses and even the hull of the boat – all made of molecules from some deep underground reservoir. On every trip out the surfers more or less heedlessly use dozens of products that without oil simply would not exist. Just as we all do, of course, every day of the week. In short, they make a perfect metaphor for our total dependence on crude oil, not only for transport but also for almost all the materials of modern life.

The tanker leaves us behind, steaming up towards the biggest concentration of refineries and petrochemical plants anywhere in the world. You should really see it by night. From a distance the impossible, floodlit tangle of three-foot-thick spaghetti that forms the basis of modern industrial chemistry takes on the twinkling profile of an entire city, with catalytic cracking towers for skyscrapers. Around the world such chemical plants consume about 10 per cent of the global oil supply – around 8 million barrels per day in 2005 – and about the same again in natural gas.[1]

It's amazing how relatively little oil makes so much *stuff*: not just surfboards but cameras, telephones and gadgets of all sorts; anti-freeze; pipes and plumbing supplies; car tyres (each contains seven gallons of oil), and asphalt to build the roads they roll on; polystyrene insulated cups; X-ray negatives, catheters, stethoscope diaphragms, oxygen tents and medical gloves; packaging (2 million tonnes a year in Britain alone); window frames; nappies; furniture; paints, dyes, inks and solvents; acrylic fibres for sweaters, acrylic resin for lenses and light fittings; PVC for raincoats and toys; plastic bottles (11 billion a year in Britain alone); food colouring, stabilizers, and antioxidants; detergents; golf balls; shoe soles and entire trainers; TVs and computers (not just the plastics but also – ironically – flame-retardant chemicals); bathtubs and shower curtains; parts for fridges, cookers and washing machines; tights; carpets; rubber gaskets, seals and hoses; plastic bags (17½ billion a year in Britain alone, 100 billion in the US); bedding; electrical cable sheathing; pharmaceuticals; adhesives; cosmetics and hygiene products; Wellingtons; paddling pools; polyurethane foam for cavity insulation; CDs and DVDs (20 billion a year in the legal market alone); rope and twine; footballs; the fleece I'm wearing now; and – my favourite, this – the chemical they sluice around the insides of wine bottles to make them shiny before the wine itself goes in.[2] In fact, most 'man-made' materials you can think of are nothing of the sort; they are 'oil-made'.

There are other ways to make plastics. The basic chemistry can equally well start from plant-derived carbohydrates as from fossil hydrocarbons, which is hardly surprising since oil is only organic matter that has been simmered in a pressure cooker for a few million years. There probably isn't a petrochemical in use today that couldn't be derived by another route from 'green' chemistry. But the question is again one of scale.

The commodities trading giant Cargill has already started to produce a renewable plastic called PLA (polylactic acid) from maize, which is now being used to make bottles, food packaging and fibres

Table 4: How much maize to replace US plastics production?

NatureWorks feedstock (maize), tonnes	320,200[a]
NatureWorks PLA output, tonnes	140,000
Percentage efficiency (output / feedstock × 100)	43.72
So to calculate necessary feedstock for a given output, multiply output by (100 / 43.72 = 2.29)	2.29
US plastics production (2005), tonnes	49,785,034[b]
Multiplied by 2.29 gives maize required, tonnes	118,931,429
US maize harvest (2005), tonnes	280,216,027[c]
Percentage of US maize harvest needed to replace US plastics production	41
US maize, soya and wheat harvest (2005), tonnes	420,155,766[d]
Percentage of US maize, soya and wheat harvest needed to replace US plastics production	27

Some figures rounded. Notes on sources and conversion factors can be found in footnote.[3]

for clothing. Cargill's NatureWorks subsidiary claims that producing bottles with PLA uses a third less fossil resources and causes 40 per cent fewer greenhouse gas emissions than using the nearest petrochemical equivalent, PET (polyethylene terephthalate). PLA also burns cleanly and is biodegradable. The new plastic is being taken up enthusiastically by companies such as Wal-Mart and Del-Monte, and PLA packaging is already being used in more than 7,000 shops around the world.

NatureWorks insists that its use of maize is no threat to food production, which is absolutely correct as things stand. But let's see what would happen if *all* US plastics were produced like this. I promise you, this is the last of these calculations.

Working at full tilt the NatureWorks plant in Nebraska could produce 140,000 tonnes of PLA per year, using 320,000 tonnes of maize as feedstock (table 4). Let's assume that all plastics could be produced from the same source with the same conversion efficiency. This is a big assumption, because nobody is yet producing other forms of plastic in this way, and this means we have to extrapolate solely from NatureWorks' numbers. Nevertheless, in 2005 the US produced 50 million tonnes of plastic from oil and gas, and it turns out that to replace this would take almost 119 million tonnes of maize, or more than 40 per cent of America's harvest that year. Even if we widen the feedstocks to include all of America's principal crops – maize, soya-beans and wheat – it would still consume over a quarter of the total.

So now we have a three-cornered fight between food, biofuels and chemicals over which sector gets the vital biomass. Since more than half the American maize harvest goes to feed livestock, it looks as if the days of the dirt-cheap burger are numbered. Already in 2006, the rapidly rising proportion of the maize crop being diverted for ethanol production in the US sent grain prices soaring, and led industry ana-lysts to predict a drop in meat production.[4] This is just a hint of the dilemmas likely to be thrown up by competition for obviously inad-equate agricultural resources during the last oil shock.

Petrochemicals may be a vital part of modern manufacturing, but how would the wider economy cope with dwindling supplies of oil? Some forecasters predict that after the peak global oil production will fall by about 3 per cent a year, which means that within ten years output will have dropped by almost a quarter, and within twenty years by almost half. Others believe the decline rate will be even steeper. This raises the question of whether the largely unbroken economic growth we have enjoyed for the last half century can continue once oil output goes into decline.

Even a cursory survey of recent economic history shows that oil is critical for economic expansion. This much is obvious because big spikes in the price of oil tend to lead to recessions. In the first oil shock of 1973 the 'Arab oil embargo' led to a quadrupling of the oil price which prompted a worldwide downturn. In the next two years US economic output dropped by 6 per cent, and unemployment doubled to 9 per cent.[5] During the second oil shock in 1979, when Iran briefly stopped exporting oil, prices more than doubled again and led to the grinding global recession of 1980–82. When Iraq invaded Kuwait in 1990, the oil price rose sharply again and tipped the world into reces-sion once more.

Since oil is so obviously important for economic growth you might have thought that economists would pay it very close attention, and might have tried to analyse its precise significance, but in fact few have done so. One exception is a group made up of Professor Alan Carruth of Kent University, Mark Hooker of the US Federal Reserve Board, and Professor Andrew Oswald of Warwick University. Together they analysed American economic data from the last fifty years and found that the oil price and interest rates were the main determinants of unemployment, and that of the two the oil price was the more important (figure 17). Their analysis showed that movements in the oil price were closely mimicked by the unemployment rate after a lag of about eighteen months, and that the link was causal. 'There are good reasons why oil shocks hurt,' argues Professor Oswald. 'When the oil sheikhs get more of the world's money, someone has to end up with less. Profits drop, firms begin to go out of business, people lose their jobs.'[6]

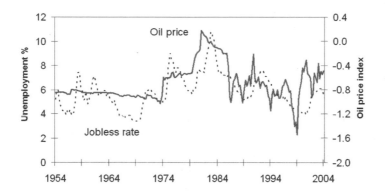

FIGURE 17. The US unemployment rate mimics movements in the oil price after a delay of about eighteen months. Source: Professor Alan Carruth[7]

Yet Professor Oswald and his colleagues are rare exceptions. To listen to many commentators you would think that energy in general, and oil in particular, scarcely mattered any more. The pervasive view among pundits and economists is that we are less dependent on oil nowadays because we use fewer barrels to produce a given amount of GDP (gross domestic product) than we used to. This statement is factually correct, as far as it goes, but the conclusion drawn is deeply misleading. The simple fact is that we use less of pretty much everything – labour, machines, minerals – per unit of output, so the relative importance of oil has not fallen, it is just part of a general increase in economic efficiency. That's what economic growth *means.*

Even if we appear to be using oil relatively more sparingly in production, at the same time each of us is actually consuming more of the stuff. Oil consumption in the Western world is falling per unit of GDP, but rising per capita, as we all drive more, fly further and consume oil in ways we hadn't previously thought of, such as buying surfboards. 'The shift in the Western economies from shipbuilding to software hardly matters,' says Professor Oswald. 'The new economy still runs on petrol and aviation fuel.'

This point was resoundingly illustrated during the British fuel duty protests in September 2000, when it was shown that the economy was if anything more reliant on fuel, not less. The severity of the crisis highlighted the fact that producing more GDP per barrel of oil may actually *increase* our vulnerability; more depends upon each barrel. Any interruption or reduction of supply has a magnified effect, spreading further and more quickly through the economy. The 'just in time' delivery ethos may have reduced business's costs dramatically, but it has also increased the fragility of the entire system to interruption in the supply of a single commodity. As the crisis in 2000 showed, once the fuel started to run short, within a week this in turn threatened employment, healthcare and food supplies. Less dependent indeed; this is the stuff without which nothing else happens.

I knew when I started to research this book that most economists didn't understand the importance of oil. But I was astonished to discover that neoclassical economics – the overwhelmingly predominant strain – still cannot even adequately explain economic growth, something of an omission you might have thought in a discipline that worships the concept. It turns out that the two errors are connected, and have profound implications for life after the last oil shock.

The story stretches back half a century, when the eminent economist Robert Solow wrote a paper outlining what is now the basis of the standard model by which neoclassical economics 'explains' growth.[8] The paper was regarded as a major advance because it debunked the prevailing assumptions about the causes of economic growth and replaced them with what seemed a far superior explanation. But it also fathered a myth that proved to be just as misleading as those it superseded. It was published in 1956 – the same year Hubbert gave his groundbreaking paper. Solow's work eventually earned him a Nobel Prize, but if you ask me, they picked the wrong guy.

In the mid-1950s economists mostly believed that the growth of economies could be explained by rising inputs from just two sources – capital and labour. Solow's advance was to show that most of the growth could *not* be explained from these inputs, and must be due to something else. He did this by taking a formula called the Cobb-Douglas production function, a formula that until then had mostly been used to analyse the relationship between capital, labour and output in individual companies, and applied it to the US economy as a whole.

Solow then introduced a critical assumption: that the relative importance of capital and labour in driving growth would be in the same proportion as the amounts of money each received in the national accounts. This in turn was based on another key assumption, that markets are efficient and that the various economic actors are therefore paid what they are worth. In general, payments to labour in the form of wages represent about 70 per cent of all the money flows in the economy, and payments to capital (in dividends, rent and so on) make up about 30 per cent. So the Solow model assumed that for every 1 per cent increase in the contribution of labour (if the workforce expanded, or workers put in more hours), overall economic output would rise 0.7 per cent, and for every 1 per cent rise in the contribution of capital (if businesses invested in more machinery) overall output would rise 0.3 per cent.

To test any model of this sort economists have to compare how well it 'predicts' the past, that is to say how closely its results match the historical record. If the consensus of the early 1950s had been correct, feeding the historical data on labour and capital inputs into the Cobb-Douglas function should have produced a 'prediction' of overall economic growth that more or less matched what had actually happened in the real world. What Solow showed was that real growth in the US economy had been vastly greater than that predicted by his

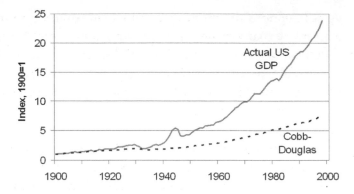

FIGURE 18. Solow's Cobb-Douglas model applied to the US economy in the twentieth century. The lower line is what the model predicts, the upper line is what actually happened. The difference is known as the 'Solow residual'. Source: Professor Robert Ayres

Cobb-Douglas model. Figure 18 shows the extent to which it underperforms reality based on US data for the twentieth century.

Quite incredibly, the question of *why* there was such an enormous gap between the model and reality seems to have troubled neither Solow himself nor the droves of neoclassical economists who followed. The clear implication was that capital and labour were far less important than the 'something else' that accounted for most economic growth, but what that something else might be was hardly debated. It was almost universally assumed that the gap represented the effects of technological advance, and that, as far as economics was concerned, was that.

The gap became known as the 'Solow residual'. Some residual! The term is usually taken to mean a minor discrepancy of no great importance, but in Solow's model the residual was the main event. In most areas of science, a model that failed to explain over three-quarters of what it sought to explain would be junked or thoroughly reworked. In economics, however, it was considered Nobel-worthy, despite the gaping question at its centre. As a result, for several decades technology remained a get-out-of-jail-free card for neoclassical growth theory, conveniently 'explaining' everything that the rest of the theory could not. Barely anyone thought to question whether the gap might have something to do with the minimal significance that the model attached to the role of energy.

It was only when a number of natural scientists started to investigate the problem that a more tenable explanation began to emerge. Prominent among them was the German physicist Professor Reiner Kümmel of the University of Würzburg. In the late 1970s, in the aftermath of the first oil shock, he was astounded to learn how little importance economics attached to energy, and resolved to put this right. 'If an influential discipline like economics disregards the first and second laws of thermodynamics, which are effectively the constitution of the universe,' he told me emphatically in July 2005, 'then it must be corrected.' Together with two respected German economists, Professor Wolfgang Eichhorn and Dr Dietmar Lindenberger, Kümmel set about devising a model that would give appropriate weight to energy and explain economic growth better.

The problem with the Solow model, Kümmel's group realized, was that the importance of energy in the economy was represented only by the amount of money spent on it. Spending on energy has traditionally been about 5 per cent of GDP, so by implication the Solow model assumed that if energy inputs increased by 1 per cent, then economic output would rise by just 0.05 per cent. But how could such a low factor remotely reflect the importance of energy generally, or even oil by itself? When the oil price spikes cause recessions, when the oil price generally determines the level of unemployment, and when cutting off the supply of fuel brings a country to its knees within the space of a week, 5 per cent is clearly an absurd under-representation of the true role of energy in the economy.

Kümmel's solution was to create a model based not only upon the traditional measures of capital (financial) and labour (hours worked), but also a *physical* measure of energy inputs used in the economy. For each economy his group studied, all the oil, gas, coal, nuclear power and hydroelectricity consumed was converted into a single number – petajoules per year. (A joule is a measure of the work necessary to raise 102 grams – the weight of a small apple – by 1 metre under earth's gravity. A petajoule is 1,000,000,000,000,000 joules.) The difficulty then was how to work out the relative weights of factors measured in unrelated units – how to combine these apples and pears.

For this, Kümmel came up with a new production function called LINEX to replace the Cobb-Douglas, which normalizes the values of the various factors of production. That is to say the quantities of capital, labour and energy were indexed to the start year, and subsequent changes measured in percentage terms to make them comparable. The LINEX function also incorporates a clever but fairly standard piece of maths that helps scientists to assess the relative importance of a number of different factors in any given outcome, when they may all be changing at the same time.

Using this approach Kümmel's group analysed the data for the US, Japan and Germany over a thirty-year period and came up with startling results. Their model produced 'predictions' of economic growth in the three countries that matched the actual outcomes very closely, almost entirely eliminating the Solow residual. This suggested that capital and labour were far less important as factors in economic growth, and rising energy consumption vastly more important. Whereas the Solow model had implied that a 1 per cent increase in energy inputs would lead to a 0.05 per cent increase in overall output, the Kümmel group's work showed that, on average, a 1 per cent increase in energy inputs accounted for economic growth of 0.45 per cent in Japan, 0.5 per cent in West Germany, and 0.54 per cent in America. So in all three, the importance of energy was about ten times greater than that implied by Solow. The results were published in a paper co-authored with the flamboyant American ecologist Professor Charles Hall, another expert in this field.[9] 'What these findings suggest,' declares Hall with his trademark bluntness, 'is that as far as energy is concerned, neoclassical economics has its head up its ass.'[10]

This is not simply a matter of academic or anatomical interest. According to Kümmel, understanding the paramount importance of energy in the economy is 'the modern-day equivalent of the revelation that the earth goes round the sun, not the other way round'. Yet most economists, and the politicians who take their advice, still believe the sun goes round the earth, and this translates into bad policy. The practical impact of this is profound: 'Because the price of energy does not relate to its productive power,' says Kümmel, 'we are wasting the most valuable resource we have on earth.'

Professor Robert Ayres took a rather different approach. Like Kümmel, Ayres trained as a physicist, but later became Professor of

Engineering and Public Policy at Carnegie-Mellon University, and then Professor of Environment and Management at the prestigious international business school INSEAD. He agreed that energy was vastly more important than the neoclassical models allowed, but wondered if there was more to the relationship between energy and growth than even Kümmel's work suggested.

With a lifelong interest in technology and engineering, Ayres was well aware of the major advances in thermodynamic efficiency that had been achieved during the twentieth century. For example, power stations in 1900 converted just 4 per cent of the potential energy they consumed as coal into usable power in the form of electricity, but by 2000 their conversion efficiency had risen almost ten-fold to 35 per cent. Substantial efficiency gains had also been made in many other areas of the economy, including high-temperature industrial processes and to a lesser degree transport.

Ayres wondered if these gains in thermodynamic efficiency could cause the economy to grow.[11] He hypothesized that advances in efficiency led to lower costs and prices, which in turn stimulated demand, boosted profits, and raised investment, so leading to fresh advances in efficiency. This kind of relationship is known as a positive feedback loop, an idea widely used in economics. But Ayres was the first to suggest that a feedback loop driven by gains in thermodynamic efficiency might be largely responsible for economic growth.

To test this hypothesis Ayres would have to use a very different measure of energy consumption; what mattered for economic growth was not total energy consumed – since so much of that was wasted – but how much was employed *productively*. It's not the coal into the power station that matters, but the electricity out. Not the petrol into the truck, but the work its rear wheels do on the road. Not the electricity into the printing press, but the amount of newspapers it spits out. So figures for gross energy consumption would have to be adjusted for the efficiency with which it was converted by machines into useful work. And those conversion rates had changed markedly over the last hundred years.

Ayres and his co-authors spent several years researching historical rates of thermodynamic efficiency in the main energy-consuming sectors – power stations, industrial processes, transport, space heating, lighting and so on – and created an index to measure *useful work* in the economy. Ayres then fed this into his model – based on Reiner Kümmel's LINEX function – producing 'forecasts' that matched actual economic growth in America and Japan for the entire twentieth century, without recalibration, almost perfectly (figure 19).

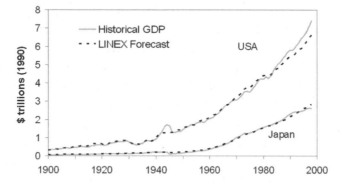

FIGURE 19. Professor Robert Ayres's model predicts economic growth in America and Japan almost perfectly for the entire twentieth century. Source: Professor Robert Ayres

There were still small residuals towards the end of the century, which Ayres speculated might represent the impact of information technology on economic growth, but in any case it was the closest fit over the longest period yet achieved. This of course doesn't 'prove' the theory is right, but there are no fudges, and no significant circularity in the analysis. I pored over the paper with Open University statistics lecturer Kevin McConway, who was impressed by its simplicity. 'The word for cheating in this kind of modelling is "over-fitted". This is not over-fitted.' The mystery of the Solow residual had been solved, and economic growth largely explained by a combination of rising energy consumption and increasing energy efficiency.

Ayres's model gives a higher correlation between energy and economic growth than any previous estimate. While the Solow model assumed a coefficient of just 0.05 per cent, and the Kümmel model found 0.54 per cent for the US, Ayres's model gives about 0.7 per cent in 2000. So by Ayres's calculation, the correlation between energy and growth is fourteen times higher than that implicitly assumed by Solow. And the higher the correlation, the bigger the trouble when the last oil shock arrives.

There have of course been brief periods in which energy consumption has shrunk and the economy continued to grow. In the wake of the first two oil crises for instance, when America made a major effort to increase energy efficiency, for eight years its oil consumption fell while GDP rose.[12] And there is clearly still a huge amount of wastage to be attacked, particularly on America's roads. But once the obvious waste has been cut, sustaining such efficiency increases relentlessly year after year will become ever harder. And in any case increasing overall efficiency takes time simply because most significant energy-consuming equipment is long-lived. A report on how to mitigate the effects of the global oil peak commissioned by the US Department of Energy concluded that to replace even half of the American car fleet would take ten to fifteen years, and more generally that 'increased efficiency alone will be neither sufficient nor timely enough to solve the problem'.[13]

Because of the thrust of his thesis, nobody is more attuned than Robert Ayres to the possibilities and limitations of efficiency gains to combat oil depletion, and he is sceptical that they could ever match the pace at which oil supplies are likely to fall after the global peak. 'The economy is *utterly* dependent on petroleum,' he says with some passion, 'and I think it is highly likely that when oil production peaks, so will the world economy. When petroleum gets more expensive everything that depends on it gets more expensive, and I cannot see how growth could really continue with much more expensive energy. It's kind of scary.'[14]

The clear implication of the new energy economics is that oil is still wildly undervalued. If energy really is between ten and fourteen times more productive than the neoclassical model allowed, then the 'real' price of oil may also be ten to fourteen times higher. I claim no authority or precision here, but since the average price of crude oil over the last thirty years has been about $24, this could imply a 'real' oil price anywhere between $240 and $340 a barrel. This may sound outlandish, but one French investment bank has in fact predicted $380 by 2015.[15] And for perspective it's also worth comparing the price of oil to that of some other liquids we value. I recently sat down in a London restaurant to notice that a 250ml glass of wine was on 'special offer' for £5. That's the equivalent of £20 a litre, or almost $6,000 a barrel. Yet wine does not drive the global economy, nor is it a finite, depletable resource.

If oil ever strays any distance into its 'real' price range it cannot stay there long because of the economic havoc it would wreak. Rising energy prices stoke both inflation and recession, so the last oil shock may mean a return to the stagflation of the 1970s, only very much worse. Soaring costs, bankruptcies and unemployment would lead to economic slump and collapsing demand for petroleum. The oil price would then subside, but given the reason, this would be no consolation. If the economy and oil demand were to recover once more, they would soon hit the oil supply ceiling again, which would be falling all the while. This process could happen many times over, or simply smear into an extended depression.

As I drive out of Houston I try to imagine what a 'real' oil price would mean for a way of life built on apparently limitless space and cheap energy. The suburbs give way to the exurbs and they in turn to the dormitory towns, connected by highways lined with car dealerships packing mile upon mile of shining metal. Daily travel in the US totalled about 4 trillion miles in 2001, meaning the average American travels 14,500 miles a year or, rather like a bad smoking habit, 40 a day.[16] Three-quarters of Americans commute alone in cars that do 24mpg on average, and in swathes of the US there is no public transport at all.[17] So what would happen here and in other energy-profligate countries such as Canada and Australia if the oil price hit even $200 a barrel? Who could afford to fill up for the journey to work, or to the mall which is also miles away? What would happen to sales of gas-guzzling SUVs, and the jobs of the people who make them? Or the price of property in the exurbs, now separated from work and food by unprecedented transport costs? Some of these factors were already beginning to bite in 2006 with oil at less than $80.

Although Europe, with much more extensive public transport, is clearly better cushioned against a soaring oil price this is easily overstated. We may travel fewer miles, but proportionately we are hardly less oil-dependent. The average Briton travels less than half the distance of their American cousin, about 6,800 miles a year, or 18 miles a day, but 80 per cent of that is by car, and only 13 per cent by public transport (and it's already overcrowded, so just imagine if everyone tried to pile on).[18] And the average commute in the two countries is not so different: about 12 miles in the US, compared with 8.5 in the UK. Either way it's a stiff walk. Besides which, the distance each of us travels – to work, to shop, for pleasure – is only half the story. The other is the distance that essential goods, especially food, travel to us.

If you pay attention to the labels in the grocery section, it's easy to see why. By offering fresh produce throughout the year, supermarkets have had to develop supply lines that stretch all over the world. Today once exotic foods go into the trolley with scarcely a thought about the distance they've come: mangetout from Kenya, avocados from South Africa, runner beans from Thailand, New Zealand chardonnay, and Guatamalan broccoli. According to a report by Sustain, a group that campaigns for locally supplied organic food, a shopping basket of twenty-six such items of produce could have travelled a total of 241,000 kilometres, the equivalent of six times round the earth, and much of it by plane.[19] To get it home would add on average another 8.3 kilometres by car.

The report also found examples of workaday vegetables being imported huge distances even during the summer when they were in season in Britain: spring onions from Mexico (8,941 kilometres by plane); potatoes from Sicily (2,448 kilometres by lorry) and onions from New Zealand (18,839 kilometres by ship). This has much to do with lower foreign labour costs, but also depends crucially on cheap transport. When the price of oil soars, much of this trade should disappear – particularly long-distance air-freight. Some sources will be substituted by local produce made more competitive by its proximity, but it is reasonable to assume that Britain will still be critically dependent on imports, even for basic foodstuffs.

Britain currently grows only 81 per cent of the potatoes it consumes, 59 per cent of the fresh vegetables, and just 9 per cent of the fresh fruit.[20] In meat we are 90 per cent self-sufficient in poultry,

84 per cent in lamb, 71 per cent in beef, and 62 per cent in pork. But even these numbers are deceptive, since European livestock depends on feed made from maize and soyabean imported from the rest of the world. It is estimated that about 3 million hectares of 'ghost land' is required to support European livestock – some of it hacked out of the Amazon jungle.[21] Overall the government calculates Britain is just 74 per cent self-sufficient in indigenous-type food, and 63 per cent in total food.[22] So on the basis of transport alone it is hard to see how we can escape a major impact on the food supply chain, whether from the soaring price of fuel, or its actual physical shortage. But food relies on oil for far more than just transport, which raises the question of whether the onset of the last oil shock threatens to undermine worldwide agricultural production, with potentially disastrous consequences.

The idea that humanity could ever outgrow its food supply has been generally held in contempt – particularly by economists – almost since the Reverend Thomas Robert Malthus wrote his famous *Essay on the Principle of Population* at the end of the eighteenth century. But at the time of publication, during a period of severe grain shortages, Malthus's ideas seemed to have an undeniable logic. He argued that the human population had the power to grow exponentially (2, 4, 8, 16), but agricultural production could only rise arithmetically (2, 4, 6, 8), and therefore 'the superior power of population cannot be checked without producing misery or vice'.[23] The amount of food per capita would tend to fall, Malthus argued, until war, pestilence and starvation restored a temporary balance, and the process started all over again.

Yet almost before Malthus's ink was dry, events conspired to show otherwise. The nineteenth century saw strong growth in both population *and* wages. At the start of the century the average European ate half a pound of meat per year, but by 1850 the figure was half a pound per week, an extraordinary rise in living standards.[24] The world population doubled during the course of the nineteenth century, and tripled again in the twentieth. So Malthus has been proved wrong over two centuries largely because he failed to anticipate the major advances in agriculture that have sustained this explosive population growth. The critical issue as the last oil shock beckons is the extent to which maintaining those advances now depends on oil and gas.

One of the most important elements in agriculture is nitrogen. All the water and sunshine in the world will not make crops grow if they lack this essential nutrient. Until the early nineteenth century the only ways farmers had of replenishing soil nitrogen was by spreading manure, or rotating their principal crop with legumes such as peas and beans, which leave behind slightly higher levels of nitrogen. While this was probably enough to maintain agricultural output at the levels of the day, it could not have supported the massive increase in yields needed to support the growing population. For that European farmers turned to fertilizer produced from the nitrogen-rich soil of the Atacama Desert in modern-day Chile, and to guano deposits from some tiny islands just off Peru. Some 16 million tonnes of this high-grade birdshit were imported to Europe, but by end of the century the Peruvian deposits had been used up, and there was growing concern about how long the Chilean fertilizer supplies would last.[25] A century after Malthus, an eminent British scientist warned that without a new source, famine would return within twenty or thirty years.

Of course there was no shortage of nitrogen; it makes up 80 per cent of the atmosphere. The trick would be to find a way to fix some of that unlimited source of the element into a form that could be absorbed by crop roots. In 1904 the German chemist Fritz Haber cracked the problem by developing a high-temperature, high-pressure technique that used hydrogen to capture atmospheric nitrogen and form ammonia. Carl Bosch, a chemist at BASF, worked out how to repeat the process on an industrial scale, and the two men later won Nobel Prizes for creating the basis of all modern nitrogen fertilizer production.

They certainly earned it; the Haber-Bosch breakthrough deferred Malthus's prediction once more by giving agriculture the means to feed a booming population through big gains in productivity. Since 1950, heavy applications of nitrogen fertilizer have raised US corn yields from 2.5 tonnes per hectare to around 8 tonnes today, and in China's most productive paddy fields, rice yields are up from 2.5 tonnes per hectare to 6 tonnes. According to Professor Vaclav Smil of

Manitoba University in Canada, roughly 40 per cent of the world's dietary protein originates from the Haber-Bosch process, and that 'without nitrogen fertilisers no more than 53% of today's population could be fed at a generally inadequate per capita level of [year] 1900 diets'.[26]

The reason why this has anything to do with oil is that the Haber-Bosch process relies overwhelmingly on natural gas – both for power and as a source of hydrogen. While gas is expected to peak some time after oil, the price of the two commodities is closely linked. So when the price of oil soars, so will the price of gas, and of fertilizer. In the developed world, at the very least food will become much more expensive. In the developing world the effect may be even more serious. Countries such as China and India subsidize fertilizer use, but governments will not be able to insulate their farmers entirely from future price rises. There may come a point when poor farmers choose not to use expensive fertilizer, particularly if they can at least feed their own families on the reduced yields. Overall production would then fall, and hunger increase.

When gas peaks and goes into terminal decline, hydrogen can still be derived from other sources such as coal – the feedstock originally used by Haber and Bosch. But the environmental cost would be even higher carbon emissions. Hydrogen can also be made by electrolysing water, but as we saw in chapter 4, this demands fantastic amounts of energy and will always be far more expensive than the traditional gas-based Haber-Bosch process. When gas becomes physically scarce there may be no alternative.

Even if agriculture could do without nitrogen fertilizer, it also depends on oil for irrigation, which is hugely energy intensive because water is heavy, and crops need a lot of it. To produce one tonne of grain takes *1,000 tonnes* of water. Luckily not all crops need irrigation in all locations, and only 17 per cent of the world's cropland is irrigated. But that 17 per cent produces 40 per cent of the world's food.[27] Three-quarters of the irrigated land is in the developing world, and in countries such as Pakistan, Nepal, Bangladesh and in northern India, most of the irrigation is powered by diesel pumps. Irrigation in the Indian subcontinent consumes some 10–11 billion litres of diesel a year. According to Dr Tushaar Shah, Principal Scientist with the

International Water Management Institute, based at Anand in Gujarat, irrigation use is sensitive to changes in the oil price: 'Energy costs are a life-and-death issue for small farmers in South Asia. My sense is that a doubling of the diesel price would have a significant impact on irrigated areas and food output.'[28]

In the West, the other big energy consumer in agriculture is machinery, which accounts for about a third of the total. Tractors and combine harvesters have replaced huge amounts of expensive human labour, largely because oil is so incredibly energy dense: a single gallon of petrol contains the energy equivalent of three weeks' human labour, at forty hours a week.[29] The result is that whereas grain production in developing countries demands 1,000 hours of labour per hectare per year, in the US one man and his machines get it done in just 10 hours.[30] In this area at least the West is far more vulnerable to the last oil shock than the developing world. When the price of oil soars and eventually – if alternatives are not available – fuel becomes physically short, none of the options is particularly encouraging.

Until the advent of mechanization, the main source of agricultural power was the horse, but this option is no longer open to us. Growing fodder for drays used to occupy up to a third of all cropland, space that is now used to grow food for humans, or feed for intensive meat production.[31] With the arrival of the tractor, oil in effect gave agriculture a massive, one-off endowment of land. And since then the human population has doubled, so we can hardly afford to give it back. If alternative fuels are not found, Western agriculture will be forced to go back to using far more human labour than it has for half a century. Here at least there may be a tarnished silver lining: if it comes to it, the last oil shock will already have created plenty of unemployment. There should be no shortage of workers to return to the fields.

So it's all very well for economists to sneer that Malthus has been continually proved wrong by human ingenuity. Ingenious we may be, but for the last century our single big idea has been petroleum, on which we now depend utterly for industrial materials, almost all our transport, and critically for food; every calorie you consume takes ten calories of fossil fuel to produce.[32] And now that our big idea is no longer big enough, we are forced to adopt the next best alternatives, which all come with stringent conditions attached. The sources that are abundant and energy dense such as coal have the potential to dev-astate the climate and life on earth. The sources that are renewable and clean are so diffuse as to make the job of replacing oil truly monu-mental. It is going to get a whole lot harder to keep proving Malthus wrong.

6

Long-term Liquidation

IN INCH-HIGH LETTERS the *Telegraph* headline screamed: 'We're Not Running Out of Oil Says Record-breaking Shell'.[1] It was the morning after Shell reported profits of almost $23 billion (£13 billion), the highest ever for a British company, and only days after Exxon had turned in astonishing net earnings of $36 billion, a global record for any company in any sector. The Shell headline reflected the two dominant themes to emerge for the industry in early 2006: the soaring oil price meant that oil companies were generating embarrassingly large profits, while at the same time being forced to rebut persistent doubts about the future availability of their stock-in-trade. The quote that supported the story was delivered by Shell's chief executive Jeroen van der Veer at a news conference in London, when I asked him when he thought worldwide oil production would peak. He replied: 'That is not how it will go. The peak oil theory itself is correct if one takes easy oil close to the market – if you look at West Texas, or even the North Sea . . . But think about deepwater, the Arctic, oil sands, shale, even coal . . . so there is not one peak. And there will be many peaks and the peaks will be on very different time-frames. And how that will develop we don't know.'

The quote was carried in several newspapers without comment, but the argument is singularly unconvincing; individual oilfields within a producing country also peak 'on very different timeframes', but as we saw in chapter 2 they still conspire to create a single overall peak. The same is true of countries within a geographical region, or when grouped together in an economic bloc. For example, total oil production in the OECD countries peaked and has been in decline since 1997, despite the fact that production in Canada and Mexico continued to grow (although Mexico seems to have peaked in 2004).

Oil production forecasting is all about calculating the cumulative effects of lots of smaller peaks that occur at different times, and this will be just as true for the various categories of hydrocarbons listed by Mr van der Veer. Despite the transparent weakness of his argument, the Shell chief executive – along with almost every other oil company boss – continues to insist that there is no global oil production peak on the horizon. But this apparent confidence begins to sound more like whistling past the graveyard when you take a look behind the enormous profits and analyse the oil companies' real predicament.

In retrospect the Shell reserves scandal was quite clearly an early warning light for the last oil shock. The company had been left behind in the wave of enormous takeovers and mergers that transformed the industry at the end of the 1990s, when Exxon and BP in particular acquired reserves and production capacity simply by snapping up other companies. Having stayed aloof from the feeding frenzy Shell increasingly found that it couldn't replace the oil it produced each year with fresh reserves by exploration alone. The *reserves replacement ratio* (RRR) is one of the most important numbers for stock market analysts when they estimate the value of an oil company, more important even than earnings; it is a basic measure of how long the company can stay in business. Any company whose RRR persistently falls below 100 per cent will find its share price falling – so threatening executive pay – and it may itself be taken over. Shell was now under enormous pressure.

As early as 1997 the company began to book as proved reserves large quantities of oil and gas that did not meet the criteria of the US Securities and Exchange Commission (SEC), and so deceived the stock market about the state of its business. The SEC defines proved reserves as those which 'geological and engineering data demonstrate with reasonable certainty to be recoverable in future years from known reservoirs under existing economic and operating conditions, i.e., prices and costs as of the date the estimate is made'.[2] Such was Shell's desperation, however, that the company broke some aspect or other of this rule from Kazakhstan to Europe and Africa to the Middle East.

In north-west Australia Shell booked as proved 550 million barrels of oil equivalent from its stake in the Gorgon gas field, despite the fact that to exploit it would mean building a massive liquefaction plant in a pristine nature reserve, and local environmentalists were doing everything they could to prevent that happening. Shell's partners in the project, ExxonMobil and Chevron, clearly thought there was no 'reasonable certainty' they would produce the gas, and booked nothing. In Nigeria the problem was that Shell's licence was drawing to a close, and any oil that could not be produced within the remaining licence period could not properly be booked as proved. As a result, the SEC concluded, Shell management cooked up 'unrealistic production forecasts that appeared to have been "reverse engineered" solely to support the reserve figures'.[3] In Oman the disparity was even starker because oil production had peaked and gone into steep decline, making Shell's claimed reserves even less credible. Two internal reviews concluded that almost 250 million barrels 'were non-compliant because they were not supported by any identified projects'.

None of this happened exactly by accident. Staff compensation was partly based on the size of reserve additions, and within Shell this was recognized as a problem long before the company came clean. As early as January 2002 the company's Group Reserves Auditor reported: 'The widespread use of reserves targets in score cards affecting variable pay is seen to affect the objectivity of staff in some [operating units] when proposing reserves additions . . .'[4] But the report's author – the sole reserves auditor for the entire company – was a powerless part-timer who was evidently browbeaten by local management, and unable to force through the de-bookings he knew were necessary.[5]

The reserves issues were discussed at the highest level within Shell for two years before the scandal broke. An investigation by the law firm Davis Polk & Wardwell uncovered documents that showed that senior management had been warned in February 2002 that more than 2 billion barrels of proved reserves were at risk because of 'non-compliance with SEC guidelines'. The lawyers also examined email correspondence between chairman Sir Philip Watts and his successor as head of exploration and production Walter van der Vijver, which showed that both men 'were alert to the differences between the information concerning reserves that had been transmitted to the public, "external", and the information known to some members of management, "internal"'. The emails make clear that van der Vijver blamed Watts for having approved unsupportable reserve estimates during his time as boss of exploration and production, while Watts pressed his successor to keep delivering 100 per cent replacement. Finally on 9 November 2003 van der Vijver exploded: 'I am becoming sick and tired about lying about the extent of our reserves issues and the downward revisions that need to be done because of far too aggressive/optimistic bookings.'[6]

Exactly two months later Shell was forced to admit to the Stock Exchange that it had overstated its reserves by almost 4 billion barrels, or 20 per cent – although this figure was later inflated by a series of further admissions. The value of Shell's shares collapsed by £3 billion in a single day, the company paid big fines to American and British regulators, and its reputation was shredded. Sir Philip Watts has always strenuously denied any wrongdoing, and a statement issued by his lawyers reads: 'Sir Philip's integrity is beyond reproach.' All charges against him, Walter van der Vijver and another executive have now been dropped.

It may have been Shell's disgrace, but the scandal had far wider implications. It highlighted the fact that big oil companies were having increasing difficulty in replacing their reserves through exploration; it showed the lengths to which some would go to persuade the market they were succeeding, and it demonstrated the kind of punishment that all could expect if they failed.

The predictably violent reaction of the stock market to the Shell scandal may also go a long way to explaining why the big oil companies are so reluctant to admit that the peak of global oil production may be imminent, or even that it will happen at all. To do so 'might suggest that oil is the next sunset industry', says Michael Rodgers, in what seems to be something of an understatement. Rodgers is a partner in PFC Energy, a Washington-based consultancy that works with all the major oil companies, and whose own detailed analysis of when global oil production will peak has been considered by the White House. 'As soon as you start talking about peak oil, you have to go the next step and explain to investors why peak oil is not really a problem for the company you're running – whether that's ExxonMobil or BP,' he told me during an interview at his home in California in September 2005. 'Look at what little soundbites do to stock prices, so I just don't think these major companies see anything to be gained from getting involved in this conversation.' They could also have a lot to lose, if Shell is anything to go by. No wonder big oil companies deny the last oil shock, particularly since their underlying position is not quite so different from Shell's as they would have you believe.

On the face of it, among the top five oil companies Shell is still uniquely bad at replacing its production, managing just 67 per cent in the five years to 2005. The other oil majors seem to be doing much better, and still generally turn in reserve replacement figures above 100 per cent. But again, when you look behind the headline numbers to analyse *how* they are achieving it, their performance is not so reassuring.

In early 2005, for instance, ExxonMobil chairman Lee Raymond proudly announced that during the previous year the company had once again more than replaced its production with fresh reserves, and it had. But not with oil. During the year the company had produced 935 million barrels of 'liquids' (crude oil and 'natural gas liquids' such as propane and butane), and 3.9 trillion cubic feet of natural gas. Since 6,000 cubic feet of gas contains the same amount of energy as a barrel of oil, for convenience many companies quite understandably express their total oil and gas production in terms of *barrels of oil equivalent* (boe). In this case, Exxon had produced 1.6 billion boe, and as it had added new reserves of 1.8 billion boe the stock market was placated.

However 1.7 billion boe came in the form of natural gas reserves in a single field in Qatar. The barrels of oil equivalent convention meant it wasn't obvious that Exxon had added only 100 million barrels of actual crude oil to its reserves during 2004 while producing well over 900 million, and that strictly speaking its oil replacement ratio was just 11 per cent. The following year the same thing happened again. Of course, there was no Shell-style deception here, since analysts understand the boe convention perfectly well, but it is a very different picture from the one suggested to the outside world by the headline figure.

Some of Exxon's competitors have been far more successful at replacing their oil production with oil reserves, but overall the situation is getting worse. And even when companies do manage to replace their production, it is important to understand how they achieve it. A report by oil analysts at Simmons & Company shows that for the fifteen biggest integrated oil companies the amount of reserve replacement achieved by *exploration* has been in decline since 1998, and that in 2004 they replaced just 60 per cent by this method.[7] As a result, companies have increasingly been forced to go 'drilling for oil on Wall Street', meaning that they replace reserves by taking over other companies. This may benefit individual players for a while, but of course unlike exploration it does not increase total global oil reserves. Such takeovers amount to shuffling the same deck of cards, a deck that shrinks as each hand is played.

Reserves are important because they give an indication of how long production can be sustained, but production itself is already under pressure among the very biggest oil companies, although again you wouldn't know it from the headline figures. The output of the so-called 'supermajors' – ExxonMobil, BP, Shell, Chevron and Total – has grown from about 8.9 million barrels a day in 1994, to about 10.5 million barrels a day in 2005.[8] This growth kept pace with the rise in global oil production, and meant that collectively the super-majors were able to maintain their market share of about an eighth of the world's daily output.

The aggregate figure conceals some marked differences however. Exxon and Shell struggled to keep production essentially flat, Chevron's output fell steeply after peaking in 1998, and all the production growth was achieved by BP and Total.[9] It is no coincidence that these are the two companies that have expanded most aggressively by acquisition. Total tripled its output after taking over PetroFina and Elf Aquitaine, while BP added well over 1 million barrels a day when it bought Amoco and ARCO in the late 1990s, and half as much again through the creation of its Russian joint venture TNK-BP in 2003. But just like reserves replacement, expanding production by acquisition does nothing to grow global output. And if you strip out this growth by acquisition, the supermajors today are producing no more oil than they did a decade ago despite the soaring oil price, despite the world market having grown 20 per cent, and despite their investing billions of dollars in their upstream operations.

The outlook for oil production from the supermajors and other big oil companies seems likely to deteriorate further. This is certainly the view of Art Smith, the chairman and chief executive of John S. Herold, one of a number of research firms that analyses production data to predict the years in which individual companies' output will peak and go into decline.[10] I meet him in September 2005, in the bar of the Houston Petroleum Club, a wood-panelled affair at the top of a downtown skyscraper, hung with portraits of its illustrious alumni. He buys me a drink and tells me how they go about it.

At any given point in time the output from an oil company's existing portfolio of fields is likely to be in decline, and that decline rate can be fairly well predicted into the future. So too can the impact of any additional production from new fields that the company plans to develop – these tend to be big engineering projects with long lead times, so the scale and timing of new production is reasonably well known for years in advance – along with date and speed of *their* subsequent declines. Then it is simply a question of calculating when the aggregate declines will start to outweigh the incremental production. 'When you put it all together you come up with projection of when each company – based on current knowledge and projects – will have a production peak.'

Smith stresses that company peaks can be delayed by takeovers. In fact that's often why they happen: 'Companies with near term peaks are often driven into the acquisition market because they're trying to patch a near term problem, it tends to drive their strategy.' And as a result peak forecasts are inevitably less hard and fast for companies than for countries. Nevertheless, based on the majors' current portfolios, John S. Herold analysts calculate that Shell's production peaked in 2006, ConocoPhillips will follow in 2008, BP in 2010, and ExxonMobil, Chevron and Total in 2011. And in a recent study of more than 200 oil companies, Smith and his analysts concluded: 'the industry continued to face the even more serious challenge of maintaining growth. Near term production has been increasing, reflecting the steady rise of capital investment over the past five years. The serious shortage of new longer term investment opportunities will make such continued growth past 2010 difficult to achieve.'[11] So it seems that while some of the fish may still be growing, their pond is drying up.

Perhaps one of the most telling clues to the oil companies' predicament is found in their financial behaviour. Since about the turn of the century the soaring oil price has meant the industry has been making literally more money than it knows what to do with, and its spending choices have been revealing. In 2005 the supermajors spent about $50 billion looking for and producing oil and gas, while returning about $76 billion to their shareholders in the form of dividends and 'share buybacks'. And by 2009 they are expected to return another $250 billion to their owners. According to financial analyst Neil McMahon of Sandford Bernstein: 'The size of the buybacks and the relatively low percentage that is being reinvested in the industry would tend to indicate that oil companies don't have enough new upstream opportunities and projects to spend their money on.'[12]

In Houston, veteran oil industry consultant Henry Groppe explains that the Western oil companies' basic problem is simple, but intractable. Groppe is a Texan contemporary of Hubbert's, and takes a similarly fundamental approach to the business of oil production forecasting. Now in his early eighties, he is still at his desk every day, and his firm Groppe Long & Littell continues to provide strategic advice to a long list of clients including big oil companies and governments. Although our interview in his modest offices on the edge of downtown Houston in September 2005 lasts several hours, he never once struggles to find a word.

For Groppe the Western oil companies' inability to replace reserves, reliance on takeovers, and massive share buyback programmes are all symptoms of the fact that they are trapped in mature and declining regions of oil production. 'The areas in which the private companies have best access to known resources are places like the US and the North Sea,' he says, flourishing a graph of the relentless production declines in those countries. 'So where's your future?' The bulk of the world's remaining reserves lie in OPEC countries, from which Western oil companies are largely excluded, except on the most restrictive terms. In these circumstances, argues Groppe, the Western oil companies are doomed to shrink further. 'The major, publicly traded oil companies', he concludes, 'are in long-term liquidation.'

Big oil companies cannot admit any such thing of course, despite the mounting evidence, but their actions betray a rather different position. Every one of them is investing in parts of the world and in forms of hydrocarbon production that they would never have considered had there been any easier or less risky alternatives. And while this diversification out of conventional crude oil may allow them to avoid or delay their extinction as companies, the big question is to what extent their investment strategies will help to mitigate the last oil shock for the rest of us.

It was ironic really, my getting a speeding ticket on the way to see oil sands production in the wilderness of Canada's Alberta province, where the critical issue is not the size of the admittedly vast reserves of this 'unconventional' oil, but the speed at which they can be produced. I was late, the highway deserted, I thought I was justified. But that's not how the company safety officer saw it when he pulled me over on the service road leading to Shell's Muskeg River site. I had to admit I was going too fast; what I had come to find out was whether the oil sands can go fast enough.

Producing oil from oil sands — or tar sands as they were once more evocatively known — is a radically different business from conventional oil production. What is produced is in fact bitumen, which one local expert described to me as 'the bottom of the hydrocarbon food chain': it's thick, heavy and sticky, and turning it into anything useful takes an awful lot of work. It is found not in deep, pressurized reservoirs, but relatively shallow deposits which range from a few metres below the topsoil to several hundred metres down. There are no drilling rigs to greet me at the Shell site, but a vast open-cast mining operation.

Everything about it is huge. The tiered pit measures almost 10,000 acres by 300 feet deep, and is crawling with what at first appear to be toy-sized dumper trucks, ferrying broken rock from diggers to the crushing unit in a non-stop relay. As we get closer the true scale becomes apparent. Each scoop of the digger lifts 100 tonnes, and each squat, massive $5 million truck receives four scoops before lumbering away, to be immediately replaced by another. To bear this 400-tonne burden their tyres are twelve feet high and five feet wide, and their engines drink around sixty gallons per hour. When the trucks dump their cargoes on to the world's largest conveyor belt, their tipper beds rise fifty feet into the sky.

The crushers process 14,000 tons of material an hour, day and night, which is then fed into an industrial-scale washing machine where hot water separates the oil from the sand in a series of massive tanks. The water and sand are sent to tailings ponds, and the sand will eventually be carted back to the hole it was dug from, and the landscape restored. But the bitumen is still too thick to flow and must be diluted with a solvent to make the 450-kilometre pipeline journey to an upgrading facility. Here hydrogen – mostly derived from natural gas – is added to the bitumen to produce synthetic crude oil. Only now is it fit to enter a refinery to be made into useful products.

'This is a very complex operation,' says Brian Straub, Shell Canada's Senior Vice-President for Oil Sands, 'much more complex than drilling conventional oil and gas wells.' It is also far more energy intensive. Two tonnes of rock must be mined to yield one barrel of oil, so the energy return is far worse than for conventional oil production. To produce one barrel of regular oil takes the equivalent of 2.5 per cent of the energy contained in that barrel, whereas a barrel of Shell's synthetic crude consumes 11 per cent – over four times as much – and the average for the oil sands region as a whole is an astonishing 33 per cent.[13] As a result the greenhouse gas emissions from production are also far worse – although Shell claims to perform better than its competitors on this count – and the costs far higher. The disadvantages are so pronounced, it makes you wonder why they bother.

One reason, of course, is the depletion of conventional oil. 'Much of the easy stuff is gone,' Straub admits candidly. 'Conventional oil and gas reserves are for the most part very mature and for the most part typically in decline. All around the world we're turning more and more to unconventionals.' By contrast, unconventional oil resources are largely untapped. According to the *Oil and Gas Journal* the Canadian oil sands contain almost 175 billion barrels of proved reserves, second only to Saudi Arabia.[14] But unlike the oil of the Middle East, here the international oil companies have open access to a huge resource in a stable country right next door to the world's biggest oil consumer. All of which has created a latterday gold-rush in the Alberta oil sands, which ironically stifles the pace at which production can expand.

As soon as the major oil companies piled into Alberta, they turned once-sleepy Fort McMurray into a boomtown. The population has doubled to 60,000 in a decade and is expected to reach 100,000 by

2010. When I was there in September 2005 the restaurants were heaving, hotel room rates outrageous, and house prices higher than in Toronto or Vancouver. But these are minor irritations compared to the rampant cost inflation besetting the oil sands companies. Shell plans to expand its output from 165,000 barrels per day to 265,000, but the estimated cost of that expansion has soared from C$4 billion to as much as C$13 billion, because of spiralling prices for steel, equipment and labour.

When I ask Straub what keeps him awake at night, there is no hesitation: 'The big three are people, people, people, it's really going to be a people challenge.'[15] Building a mining and upgrading operation is a huge engineering endeavour, he explains, and it demands thousands of skilled workers – chemists, geologists, welders, pipe-fitters and carpenters – for which the industry is already having to scour the world. When I interviewed him, Straub had just been on a recruiting trip to Venezuela and Brazil, and was also expecting to hire from as far afield as eastern Europe and Asia. In total he estimates the industry in Alberta will need another 40,000 workers between 2010 and 2015, and 'we are clearly concerned we may not get our fair share of them.' Some oil sands projects have already been delayed for several years – and at least one cancelled altogether – by fierce competition between operators for staff and materials, and Straub expects this to continue: 'Not all of these projects that are currently on the map will happen; we do have a resource limitation called people.'

Oil sands production doesn't only require thousands of skilled people, it also consumes huge amounts of water and natural gas, and these too are likely to inhibit its growth. Natural gas is needed not only to make hydrogen to upgrade the bitumen, but also for the extraction process itself. Mining operations burn large amounts of gas to generate heat and power, but there is another method of oil sands production that uses even more. Only 20 per cent of the oil sands reserves are shallow enough to be mined, and the rest has to be produced *in situ*, often using a technique known as Steam Assisted Gravity Drainage. This involves drilling two horizontal wells, one above the other, and pumping steam into the upper well to melt the bitumen around it, which then drains downwards to be collected by the lower well. And on average it takes 1,000 cubic feet of gas to generate enough steam to produce one barrel of bitumen.[16] Since four-fifths of the oil sands will have to be produced using this technique, the industry's natural gas needs will soar as production grows.

In a strategy paper published in 2004, the Alberta Chamber of Resources proclaimed a 'vision' that oil sands production should reach 5 million barrels per day by 2030.[17] But the report goes on to explain that to achieve this would consume an 'unthinkable' 60 per cent of all the gas available in western Canada each year. According to Len Flint, the British-born consultant who wrote the document, 'This is totally unsustainable. We wouldn't have society on our side, and besides which it's rather like turning gold into lead.'[18]

There are alternatives to natural gas however. Several companies are beginning to design projects which in theory could liberate oil sands production entirely from its dependence on natural gas. By taking the heaviest fraction of the bitumen and mixing it with oxygen and steam, it is possible to produce a synthetic gas which can then be used to provide both power and hydrogen to upgrade the remainder. Using 20–30 per cent of the recovered bitumen in this way could make the whole process self-sufficient. This would be an elegant solution were it not for the fact that it makes oil sands production even filthier in terms of greenhouse gas emissions than it already is. But the technology is still in its infancy and cannot be entirely convincing since the French oil giant Total is seriously considering doing the same job by building a nuclear power station.

Water may be an even greater problem than natural gas. The mining operations in Alberta use between three and ten barrels of water for every barrel of oil produced, and although much of it is recycled the industry consumes as much as 10 per cent of the flow of the Athabasca River.[19] Over lunch in Calgary, Len Bolger, a retired senior Shell executive and the chairman of the Alberta Energy Research Institute, tells me that the current water supply will support a maximum oil sands production of 3 million barrels per day. 'There's nothing in sight which will solve the water issue,' says Bolger, 'it's a huge problem.'

Given all the obstacles, many observers think it highly unlikely that Alberta will achieve its oil sands 'vision' by 2030. The locals are naturally more optimistic, but still cautious. For Len Bolger, 'It's do-able, but we can't get there with today's technology.' Shell's Brian Straub sounds doubtful too: 'Five million barrels a day would be a substantial achievement for this industry, especially relative to the resource limitations that we have.' But even if Alberta does hit its target or even exceeds it by a handsome margin, conventional oil production, depending on the year in which it peaks, could already have fallen by well over 30 million barrels a day. As the Americans have it, do the math.

As a honeypot for oil companies, Alberta is matched only by Qatar, the thumb-like peninsula poking north from Saudi Arabia into the Persian Gulf. Its oil reserves and production are modest, but its gas resources are huge, and unlike its neighbours Qatar has welcomed foreign oil companies to help exploit them. This is where ExxonMobil has been 'replacing' its oil reserves by developing several massive gas projects. Others such as Shell and ConocoPhillips are doing the same, but Exxon has by far the biggest position. A further queue of supplicants is waiting at the door.

The tiny desert state has a population of just 730,000, three-quarters of whom are foreign guest workers, and hosts a massive and secretive US military base south of the capital Doha, which served as the command centre for the invasion of Iraq. Now it stands guard over the country with the third largest gas reserves, contained in the world's single biggest field. The North Field makes up the greater part of a 6,000 square kilometre, predominantly offshore geological structure which straddles the north coast of Qatar, and stretches across the maritime border towards Iran. The Iranian section is called South Pars, and taken as a whole the field is four times larger than the world's next biggest gas field, Urengoyskoye in Russia. Between them Qatar and Iran hold one-third of global proved gas reserves.[20] No wonder the US base is where it is.

At the tip of the peninsula, Ras Laffan is the huge industrial complex where the North Field gas comes ashore. The trip up from the capital – half the length of the country – takes only an hour, through flat, hard desert, dun-coloured except for a single flash of emerald, the irrigated greens of the Doha golf club. The first sign to emerge through the thick Gulf haze that we are approaching Ras Laffan is a forest of electricity pylons, followed by two disembodied flames dancing high in the sky like genies, an impression only dispelled when we get close enough to make out the 150-metre flare stacks from which they emanate. At their base is a rapidly expanding city of convoluted pipework and processing plants which will soon make Qatar the world's biggest exporter of liquefied natural gas (LNG) and gas-to-liquids (GTL) fuels. Again, the key question is to what extent these developments, and the expansion of natural gas production globally, might mitigate the last oil shock.

Gas-to-liquids is a way of producing synthetic liquid fuels from natural gas, using the Fischer-Tropsch process developed by two German researchers in the 1920s. The process combines carbon monoxide and hydrogen to form a synthetic gas, which is then put through a catalytic chemical reaction to produce diesel or jet kerosene. But as we saw in chapter 4, the hydrogen has to come from somewhere. The original Fischer-Tropsch feedstock was coal, which Britain used to make 'town gas', Germany to fuel its armies in the Second World War, and South Africa to beat the international oil embargo during the apartheid years. But wood or gas will also do.

Since remaining reserves of coal and gas are often said to be far larger than of oil, you might have thought that Fischer-Tropsch fuel production could provide at least a temporary respite from the last oil shock, but as usual there are major drawbacks. In the case of coal-to-liquids (CTL) fuels the greenhouse gas emissions are twice as high as those from conventional diesel.[21] But that's done nothing to stop a major expansion of coal-to-liquids production being planned in America, India and China, all of which have massive reserves of coal. If oil shortage leads to unfettered expansion of CTL, the impact on the climate could clearly be catastrophic.

In the case of GTL the greenhouse gas emissions are lower than for CTL because the feedstock is so much cleaner, but still higher than for conventional diesel. This is because the Fischer-Tropsch process is very energy intensive. According to the International Energy Agency, gas-to-liquids production is just 55 per cent efficient, meaning that the production process itself consumes 45 per cent of the gas.[22] This and the high capital costs of GTL plants also make the process extremely expensive. Shell's Pearl project is planned to produce just 140,000 barrels of GTL products per day, and about the same amount in by-products, yet the estimated cost of building it has soared to between $12 and $18 billion. If everything stays on schedule, which is a big if, the IEA forecasts total Qatari GTL production of some 300,000 barrels per day in 2011, and that global GTL output will reach just 2.3 million barrels per day by 2030.[23] All in all, not much of a lifeboat.

Natural gas does not have to be chemically transformed into a synthetic liquid to serve as a transport fuel however. Existing vehicles can fairly easily be converted to run on compressed natural gas (CNG). And an authoritative 'well-to-wheels' study conducted for the European Commission concludes that the hybrid CNG vehicles which should be available from about 2010 will have better overall energy return and greenhouse gas emissions than vehicles running on conventional fuels.[24] So in theory falling supplies of crude oil could be replaced by natural gas without too much problem: a new refuelling infrastructure would be needed, and you would lose some boot space, but it could be done. The remaining questions are whether the natural gas resource is sufficient, and whether production can be expanded quickly enough both globally and within individual regions.

The range of estimates for the world's ultimately recoverable natural gas – including the gas already consumed and yet to be found – is wide. Of about thirty assessments of the global resource conducted since 1980, the lowest was smaller than 8,000 trillion cubic feet, and the largest more than 20,000 trillion cubic feet.[25] The mean average of those estimates is 11,700 trillion cubic feet, which by way of comparison is the energy equivalent of about 1.9 trillion barrels of oil. Jean Laherrère, the former head of exploration technique at Total, uses a Hubbert-style logistic curve analysis of natural gas production (see chapter 2) to derive an ultimate of 12,000 trillion cubic feet, which coincidentally happens to be the same number Hubbert produced for the USGS as long ago as 1973.[26]

It is not clear how much of the total gas resource has been consumed so far, because for many years when oil and gas were discovered in the same reservoir, the 'associated gas' was often simply flared off. This shocking waste still happens in places such as Nigeria and Russia. But despite the uncertainties it seems clear that we have used a much smaller proportion of the gas than the oil. Jean Laherrère estimates cumulative production to the end of 2005 was 2,860 trillion cubic feet, about a quarter of the gas ultimate, as opposed to much nearer half for most sensible estimates for oil. So on the face of it there seems to be enough to be getting on with to support the use of gas as an interim transport fuel, particularly since getting the gas to market from remote locations has been made easier by the advent of liquefied natural gas.

As its names suggests, LNG also involves turning natural gas into a liquid, but only because it has been cooled to minus 160°C. Unlike GTL there is no chemical transformation involved, and when LNG is allowed to warm up again, it returns to its gaseous state. At Ras Laffan, as we drive along one of the LNG 'trains', so called because the complex of cooling units and gleaming pipework is arranged in a straight line over more than a kilometre, my guide explains that it works just like a domestic refrigerator only bigger. As a liquid the LNG occupies only 1/600th of the volume of natural gas, so a single tanker-load carries enough gas to supply over 60,000 British homes for a year.[27] The tankers run their engines on the relatively small amount of gas that evaporates from their cargo, which can be delivered to regassification terminals anywhere in the world.

No country has bigger plans to develop LNG than Qatar, as His Excellency Abdullah bin Hamed al Attiya, second Deputy Prime Minister, Minister of Energy and Industry, and chairman and managing director of Qatar Petroleum, was proud to explain. Sipping tea in his grand office in Doha, wearing the traditional white robe and headdress, the minister recited a long list of LNG projects involving ExxonMobil, Shell and other international oil companies, and a series of targets he was confident his country would soon achieve: Qatar would overtake Indonesia as the world's largest exporter; Qatar would be the only producer to export to Asia, Europe and North America simultaneously; and by 2012 Qatar would be producing 75 million tonnes of LNG per year. His ruddy face beamed as he declared, 'We will be the biggest in LNG!'

Nobody much doubts that Qatar will pass these milestones, but even this country's legendary gas resources now seem to have limits. During our interview in April 2005 al Attiya hinted that the furious pace of development was not sustainable — 'we are taking what you call a breather' – and within weeks he had announced a moratorium on new gas projects. At first it seemed this was just to relieve engineering bottlenecks of the sort being encountered by oil sands companies in Canada, but it soon emerged that there were also more fundamental concerns about the effect of such rapid expansion on the state of Qatar's gas reservoirs.

In June 2005 al Attiya told a conference in London that Qatar was now focused on treating the North Field 'very carefully . . . to be on the safe side', to allow the country '100 years of production, not 25'. Under the headline 'Qatar Seeks New Math for North Field', the authoritative industry newsletter *World Gas Intelligence* raised an eyebrow at some of the minister's remarks:

> 'We want to make sure we have booked reserves', the minister said – a slightly surprising statement given that official reserves estimates for the North Field were tripled to 900 trillion cubic feet as recently as 2002. A source at state Qatar Petroleum explains: 'We are using a little caution on how the North Field is exploited. We are already looking at a huge increase in production of gas. We are just saying, let's take it easy on the rate of development until we know more about the geology.'[28]

For almost a year the immediate cause of this sudden attack of caution was not clear. But then in April 2006 oil analyst Robert Kessler of Simmons & Company reported an industry rumour that ConocoPhillips had drilled a 'dry hole' in Qatar's North Field, and that this had prompted the development moratorium.[29] The company itself refused to comment, but the story seems well sourced and Kessler's analysis of ConocoPhillips' financial reports also supports it. If true, this might suggest that the North Field is not a single homogenous structure, and would therefore raise doubts over the size of its reserves, and the potential to increase production in future. All of Qatar's planned LNG output is already committed for the next twenty-five years, with a queue of additional buyers waiting to be allocated gas, and now it looks as if the country's gas production may grow far more slowly from now on, if at all. Al Attiya himself told me: 'Until 2012 we will see ourselves very, very busy. Beyond that, we will see.'

Qatar is not the only country where gas production is likely to disappoint. While the investment boom in LNG continues the world over, these massive engineering projects have suffered delays and cost overruns. Shell now expects its Sakhalin II project in the Russian far east to cost $20 billion – double its original budget. As a result, according to analysis from Deutche Bank Securities, LNG supply has persistently failed to meet forecasts, and will continue to do so. In a detailed report entitled *Global LNG: Waiting for the Cavalry*, analyst Paul Sankey noted: 'although there seems to be abundant supply of LNG relative to demand, in fact project delivery has been relatively poor and the market has stayed tight despite the seeming excess of gas reserves.'[30] Having considered the slate of major projects that are now being built or planned, and the industry's past record in delivering, Sankey concluded: 'LNG was seen as "the cavalry" coming to save the day. In reality we are still waiting. . . Our conclusion is that LNG supply will stay tight for the foreseeable future, being 2015 and beyond.'

Taking a longer perspective, the International Energy Agency forecasts global gas demand will rise by more than two thirds by 2030, driven largely by soaring demand for power generation, particularly in countries such as China and India.[31] The Agency argues that because of declining gas production in America and Europe, much of the additional supply will have to come in the form of LNG from Middle East and North African countries. The IEA says that to meet projected demand their exports need to quadruple by 2030, but it clearly doubts that they can do it: 'A critical uncertainty is whether the substantial investments needed in the upstream hydrocarbons section in MENA countries will, in fact, be forthcoming.' If global gas supply cannot even keep up with the growth of traditional sources of demand such as power generation, there is almost no hope of it also replacing declining oil supplies as transport fuel – at least not without equally damaging consequences in other sectors of the world economy. There may be a sizeable resource left to exploit, and there may be a boom in LNG investment, but the gas is already spoken for.

The Shell scandal is not the only evidence of high-level desperation. While some supermajors seek their salvation in gas and unconventional oil, others have maintained their focus on conventional crude, but have done so by pouring billions of dollars into some of the riskiest parts of the world, as BP did when it announced a huge Russian joint venture in 2003. The deal looked good on paper: reserves and production were increased substantially, and BP had beaten its international competitors to establish a bridgehead in one of the few remaining non-OPEC countries showing any significant production growth. However, this was Lord Browne's second foray into what is known in investment circles as the 'wild east', and the story of his earlier adventure highlights the enormity of his latest gamble.

The first deal was signed in 1997 in Tony Blair's office, where BP stumped up $571 million for a 10 per cent stake in a Siberian oil producer called Sidanco. BP seemed confident, blithely talking up the gains that would flow from the application of 'a little bit of BP management expertise to such a huge resource base', and praising the 'openness' of Sidanco's management.[32] But it soon became clear that BP's investment approach had been anything but expert, and Sidanco far from open. As the *Financial Times'* Moscow bureau chief Andrew Jack relates in his book *Inside Putin's Russia*, like most Russian oil companies at the time Sidanco had three sets of accounts: 'one for a handful of insiders, another for its partners and a third even less flattering for the tax inspectors'. It turned out that Sidanco had been systematically stripping cash out of its main operating subsidiary and prize asset, Chernogorneft, which was soon declared insolvent. That's when things began to get interesting.

BP had walked, apparently unwittingly, straight into the crossfire of a bitter feud between two 'oligarchs' who had made their billions in the Russian privatization deals of the mid-1990s. Vladimir Potanin, from whom BP had bought its stake in Sidanco, and Mikhail Fridman, who controlled the financial group Alfa, had previously fallen out over a deal involving Sidanco shares. Fridman felt betrayed and was determined to seize control of Chernogorneft through his own oil company TNK. He appointed a wily former professional chess player called Simon Kukes as chief executive to see it through.

TNK manipulated the Chernogorneft insolvency ruthlessly from the start, buying up the debts held by creditor companies in order to win control of the committee that would oversee the process. This allowed TNK to spin things out for so long that Sidanco also collapsed, forcing BP to write off $200 million. With its rival out of the way, TNK was then able to engineer control of Chernogorneft itself.

When it looked as if BP was going to lose everything, Lord Browne persuaded Blair to write to Putin, and lobbied Washington to block a $500 million American loan guarantee to TNK.[33] In the end the dispute was resolved in what looked like check-mate to Kukes: TNK secured a 25 per cent stake in Sidanco in return for giving back Chernogorneft. BP later felt compelled to protect its position by raising its stake to match TNK's, at a cost of another $375 million. In short, Lord Browne had been shafted.

When Andrew Jack asked a senior BP executive in 1999 why the company didn't simply open peace talks with TNK, he got the reply: 'You don't talk to someone who's stolen your wallet.' And yet four years later, BP was not only talking to these self-same oligarchs, but entrusting them with a further $7 *billion*. In an extraordinary about-turn, BP's new joint venture in 2003 was with TNK, still controlled by Mikhail Fridman and his financial partners, the very people who had 'stolen BP's wallet'. During the course of the negotiations, Lord Browne even felt compelled to ask the Russians, 'Are you going to take the money and run?' and yet still he signed the contract.[34]

Delving into BP's production figures before and after the deal exposes the predicament the company faced. In the early years of the century BP had disappointed the stock market by repeatedly missing a series of production targets, and had struggled to produce 2 million barrels per day. After sealing the TNK deal in late 2003, BP's production jumped to over 2.5 million barrels per day, and growth in its Russian output since then has offset steep declines at its core operations in the UK North Sea, Europe and America. Had Lord Browne flinched and *not* done the deal, however, the figures show that BP's production would have slumped to scarcely more than 1.6 million barrels per day, a rather harder story to explain to shareholders.[35] It looks as if Lord Browne had very little choice.

The fact that Lord Browne has taken such enormous risks may call into question his persistent denials that worldwide oil production will peak some day soon, even among his biggest fans. Robin West, a former US Assistant Secretary of the Interior who went on to found the specialist consultancy PFC Energy, has great admiration for BP's chief executive, but argues that the TNK deal is telling. 'John is clearly one of the most talented people in the industry, he's a really brilliant man. But I think his actions belie his words, because he's taken some big risks – as in Russia and TNK – to position himself with access to resources, the way others haven't. There's enormous political risk. There's risk with his partners and risk with the government, but he has worked assiduously to try to manage that.'[36]

Lord Browne's gamble may be paying off for BP – for the moment – but the outlook for Russian oil production more generally is darkening. The strong growth in output seen since the turn of the century has recently suffered a sharp slowdown, and may soon stop altogether.

After the fall of communism Russian oil production collapsed from a peak of almost 11.5 million barrels per day in 1987 to a low of just 6 million in 1998. Much of the industry was sold off, but without a reliable legal framework its new owners were nervous of investing more capital, and fields were mothballed or simply not maintained. According to Stephen O'Sullivan, an oil analyst with the Moscow-based stockbrokers Deutsche UFG, 'If something broke down, nobody fixed it,'[37] so output inevitably slumped.

With Putin's election as President came a new rapprochement between the state and the oligarchs, after which investment resumed and production soared, rising 50 per cent between 1999 and 2004.[38] Just as well it did: the additional 3 million barrels per day of Russian production single-handedly saved non-OPEC supply from peaking and going into decline during that period, and allowed the world narrowly to escape a real energy crunch in 2005 when the hurricanes hit Louisiana. Had it not been for Russian growth, non-OPEC supply would have fallen by 2 million barrels per day between 1999 and 2004, as opposed to rising by just under 1 million barrels per day. Speaking at a conference in London in February 2006, commodities analyst Kevin Norrish of Barclays Capital remarked, 'In retrospect, if we hadn't had that surge in Russian exports, God only knows where the price would be today.'

However, as Norrish went on to say, 'The scary thing is how that growth has tailed off very, very quickly.' Whereas Russian output leapt by 11 per cent in 2004, the rate of increase collapsed to just 2.7 per cent the following year. This was partly the result of the fear that spread among investors when Putin dismembered and effectively renationalized Yukos, one of Russia's biggest and most dynamic oil companies, and jailed its politically threatening oligarch Mikhail Khodorkovsky. But more fundamentally it was because the industry had run out of easy growth opportunities. Oil companies could no longer rely on simply refurbishing existing fields that had previously been allowed to decline, and were having to develop entirely new fields, a process that is far more expensive and time-consuming. In future, growth will also increasingly come from largely untapped provinces such as East Siberia and the Arctic continental shelf where working conditions are especially hostile. According to Stephen O'Sullivan, 'The quick wins have already been achieved.' Norrish concurs: 'It gets very much harder from now on.'

As a result a second Russian oil peak, lower than the Soviet high, is now in sight, and not simply for peak oil enthusiasts. In late 2004 the Russian energy minister Viktor Khristenko announced that his country's production would reach a plateau of just over 10 million barrels per day by about 2010, barely 1 million higher than in 2006.[39] Stephen O'Sullivan's field-by-field analysis comes up with roughly the same forecast, and others are yet more bearish. It seems that this cavalry brigade has already been and gone.

Almost every aspect of the big oil companies' behaviour belies their confident public pronouncements and betrays their real predicament. But one facet of the problem that they do acknowledge openly is their exclusion from the Middle East, where most of the remaining reserves are concentrated. At a conference in London in February 2006, Paolo Scaroni, the chief executive of the Italian oil company ENI, reflected on the fact that the international oil companies control far less than 10 per cent of conventional global oil reserves, and that the hydrocarbons they could get at were increasingly difficult to produce. 'The confluence of all these factors explains why the international oil companies are living in a very peculiar paradox of plenty: they are awash with enormous cash flow, but their opportunities to reinvest that cash are severely limited.' Thierry Desmarest, the chief executive of Total, has also admitted that without greater access to OPEC resources, his company – and by extension the industry – will not be able to increase production.[40] Total's head of exploration and production, Christophe de Margerie, has also been disarmingly frank. At a news conference in Calgary held to discuss Total's investment in the Alberta oil sands, de Margery made clear that the company's real hunger was for Middle East reserves: 'They don't belong to us, at least not yet. It's the next target for all of us.'[41]

Perhaps by talking openly about this part of their predicament, the oil companies hope to exert pressure to relieve it. No doubt they are lobbying furiously behind the scenes. However, previous attempts to open up the Middle East to Western capital have not been notably successful. Iraq's oil production languishes well below its pre-invasion level and is likely to stay there; Western companies will keep away for as long as the butchery continues. And there is no reason to think that the other Gulf producers will voluntarily throw open their reserves to foreigners, because most have their own technically competent state-owned oil companies to do the job. The real question is not so much whether international oil companies secure significantly greater access to OPEC reserves, which seems unlikely, but how long the organization itself can continue to increase its overall production. Many serious forecasters predict that total oil production for the world *excluding* OPEC will peak and go into plateau from about 2010. So from that point onwards, everything depends on the cartel.

7

The Riddle of the OPEC Sands

THE OIL MINISTERS must love the power trip. Whenever OPEC convenes to set production quotas for the coming months the meeting attracts a caravan of news agency reporters and business correspondents from around the world. For forty-eight hours scores of journalists scheme and scramble to catch the delegates between negotiations, and to be the first to report every ministerial word. This cut-throat competition traditionally climaxes on the afternoon of the second day with a ritual humiliation of the journalists charmingly known as the 'gang-bang'.

The sign perched on a brass stand in the lobby of the Intercontinental Hotel in Vienna reading 'OPEC Press Corner' was superfluous by the time I arrived one freezing Saturday at the end of January 2005. There was no mistaking the profession of the people idling watchfully among the laptops and dead coffee cups. The over-flowing ashtrays suggested they had been there for some time. 'We're waiting for Iran and they're late,' explained Reuters' energy correspondent Barbara Lewis, with one eye on the main entrance. Everybody twitched as a pony-tailed television cameraman headed outside, then relaxed a little as he turned, grinning, and held up a cig-arette to explain his intention. But journalists began to gravitate towards the revolving door anyway, just in case.

When the Iranian oil minister Bijan Zanganeh finally walked in, I glimpsed him for only a second before he was engulfed by a scrum of reporters shoving tape recorders in his face and lobbing questions. TV crews held their twenty-pound cameras precariously at arm's length over the heads of the rest, and flashlights popped like automatic fire. The scrum crabbed its way slowly across the lobby as journalists with Dictaphones in one hand and mobiles in the other relayed each vital phrase back to their newsdesks, and fur-clad Viennese ladies looked on bewildered. By the time the minister made it to the lifts, the whole crush had turned 180 degrees in unison and tipped over the OPEC Press Corner sign in the process, not that anybody noticed. By the time the minister made it back to his suite the snap headline on the Reuters newswire read: 'Iran Worried About Level of OPEC Compliance on Quotas'.

After another two hours' hanging around, a similar mobbing of the Nigerian Presidential Adviser on Petroleum, Edmund Daukoru, yielded: 'Nigeria Says Not Too Much Worried About Q2 Oil Demand'. To the civilian such esoteric comments may not sound desperately important, but any hint of a change in the future level of OPEC oil production is critical for the oil price. 'One word from a minister can move the market,' Barbara Lewis explained, 'and there's a lot of money at stake.' OPEC members do, after all, account for some 40 per cent of daily oil production, and claim to control three-quarters of the world's remaining proved reserves.

Journalists may tease out the gist of the various OPEC members' negotiating positions during the course of the proceedings, but the final decision about the overall production ceiling never emerges until the second day. And that's where the gang-bang comes in. After the main meeting the conference room is thrown open to all journalists for a fifteen-minute 'open session' when anyone can approach any minister for a quote. At the OPEC headquarters in Vienna, the conference room is four flights up from the press centre, and when the word comes the entire press corps stampedes up the stairs, pushing and shoving, since only those with the sharpest elbows have any chance of success. 'Stick very close to the stairs and wear running shoes,' is the advice from one female gang-bang veteran. 'The first time I came I wore high heels and almost got trampled to death.'

I suspect that some of the more macho journalists believe that they are the ones doing the banging – it's their phrase, after all – but in fact the opposite is true. The entire performance seems designed to demonstrate the power of OPEC, and serves as an apt metaphor for its relationship with the West. I was looking forward to seeing it for myself. So imagine my disappointment when, for the first time anyone could remember, the gang-bang was cancelled. For some unexplained reason OPEC officials opted for a conventional, orderly news conference downstairs instead. But this too may be a sign of the times: OPEC's role is changing.

OPEC's power and fearsome reputation were established in 1973 when it imposed the 'Arab oil embargo' on allies of Israel during the Yom Kippur War, causing widespread panic and a worldwide shortage of oil. OPEC had tried to impose such an embargo once before, during the Six Day War in 1967, but had failed to make much of a dent in the world's oil supply. This time the strategy worked a treat because US production had peaked in 1970, as predicted by Hubbert, and the Western world was fresh out of spare capacity. OPEC was mighty because the oil market was tight, although this turned out to be a temporary and self-correcting condition.

The two oil shocks of the 1970s – the 'Arab oil embargo', and the Iranian revolution in 1979 – sent the price of oil skywards. In 1970 a barrel cost less than $2; at the end of the decade it was flirting with $40. But by the early 1980s, price and panic had turned shortage into surplus: a deep global recession had depressed demand for oil; industrialized countries had massively improved their energy efficiency; and the new oil-producing provinces of Alaska, Mexico and the North Sea were now yielding millions of barrels a day. It all added up to a worldwide glut, in which the new production undercut OPEC's official 'posted' prices, usurped a large part of its market share, and slashed the revenues of its member countries. It was a defining crisis for the organization, and in 1982 members were forced to introduce a system of quotas to limit their own production to try to support the oil price. By doing so OPEC adopted the role it has been stuck with ever since – that of 'swing producer'.

What this meant was that OPEC, despite its massive reserves and idle production capacity, would no longer compete for market share. Instead it would limit its oil output so as only to make up the difference between whatever the rest of the world could produce and global demand. From now on OPEC would act as a concertina. If non-OPEC supply grew, or if global demand fell, OPEC would be squeezed. If demand grew faster than non-OPEC production, OPEC output would expand – although this seemed only a remote possibility when the quotas were introduced. In both sets of circumstances, non–OPEC oil producers would take as much of the global market as they could, and OPEC would make do with whatever was left.

Not that they wanted to, of course. 'The idea that somehow OPEC is a swing producer is the stupidest thing I ever heard,' exclaims Sadad al-Huseini during a lengthy interview in Bahrain in April 2005. Until the previous year al-Huseini had been the Executive Vice-President in charge of exploration and production at Saudi Aramco, effectively second-in-command at the world's most prolific oil company. Saudi government officials had ignored all my requests for a visa, so he drove across the twenty-six-kilometre King Fahd Causeway from the mainland to meet me. 'What does it mean to be a swing producer? I mean, is that to accommodate idiotic investments [in the rest of the world]?'

But the OPEC countries didn't have much choice. The commercially driven international oil companies – such as Exxon, Shell, BP and Total – with their quarterly earnings targets, and shareholders on their backs, had every incentive to develop and sell as much oil as they could, almost *regardless* of the wider picture. 'I always compare them to a herd of buffalos,' says al-Huseini in flawless English. 'Whenever they discover an oilfield anywhere in the world, they're in such a rush to develop it that they really have no concept of global supply and demand.' But if OPEC countries – with their vast oil resources – did the same, the price could only collapse. It was a game of chicken, and OPEC had more to lose.

OPEC was now left with a massive overhang of involuntary spare capacity – fields that had been developed at great expense with the necessary wells and pipelines, but were now idle or 'shut in' – and most of it was in Saudi Arabia. In 1985 Saudi production slumped to just 3.6 million barrels per day, although only four years previously the kingdom had been producing more than 10 million barrels per day, and it still had the wherewithal to do so.[1] As a result al-Huseini spent several years firing staff and mothballing facilities. 'That was quite an education,' he remarks drily. 'We don't need to repeat that.'

OPEC's spare capacity declined gradually as the world economy grew, but the organization always seemed to have enough production held back to come through in a crisis. When Iraq invaded Kuwait in 1990, the international embargo on the two countries instantly deprived the oil market of 4 million barrels a day, and it looked as if a third oil shock was in the offing. As a cub financial reporter at the time, I remember interviewing the former Saudi oil minister Sheikh Ahmed Zaki Yamani in London on the eve of war and getting very excited about his prediction of $100 oil. It never happened, because the oil market deficit was quickly replaced from OPEC spare capacity: Saudi Arabia alone increased its production by 3 million barrels a day. As it turned out, $40 oil was enough to tip the world into recession again, but there was no outright physical shortage.

Events such as this encouraged the belief that OPEC could always be relied upon to make up the difference between non-OPEC supply and global demand, whatever the circumstances. Economists, oil analysts and government officials all succumbed to the reassuring view that the 'call on OPEC' could expand almost indefinitely, and of course it suited OPEC to let them believe it; the idea that the organization effectively possessed a bottomless well would minimize any incentive for its customers to develop alternatives to oil. But in the early years of this century OPEC's shrinking pool of spare capacity finally evaporated in the heat of unexpectedly strong demand growth, and as large chunks of non-OPEC production went into plateau or outright decline. By the time of the Vienna conference I attended in early 2005, it wasn't just the gang-bang that had disappeared; so had OPEC's ability to open the spigots on demand. OPEC was no longer the swing producer, but pumping full-bore, just like everybody else.

Yet the view persists that OPEC can continue to plug the gap between failing non-OPEC supply and growing demand for decades to come. Production forecasts from Shell, Exxon, the International Energy Agency (IEA) and the US Department of Energy's statistical arm (the Energy Information Administration, or EIA), all assume that most demand growth from now on will be satisfied by a massive increase in OPEC production. In the IEA's reference scenario, for instance, OPEC will raise production by almost 70 per cent to more than 56 million barrels per day in 2030, when Saudi Arabia alone is expected to be knocking out an astonishing 17.6 million barrels per day.[2]

All these forecasts – and their implication that the global peak can be put off for decades – rely entirely on the assumption that OPEC countries are both willing and able to increase their oil production greatly from now on. Since these states claim to possess three-quarters of the world's proved reserves it is an easy assumption to fall for. But the figures, while astronomical, may be deceptive, and in fact officials in some OPEC countries are beginning to semaphore that they are not capable of satisfying unfettered global oil demand. In Vienna the then Libyan Prime Minister Shokri Ghanem told me during a snatched interview that OPEC was already pumping flat out and might not be able to keep up with demand growth in future. In any event, the world would never produce much more than 100 million barrels per day, he said, and output would decline after that peak. 'There could be an energy crunch,' he admitted. But if OPEC still has all this oil, how come?

The publicly known 'facts' about OPEC are these: its eleven members have proved reserves of more than 900 billion barrels of oil, 75 per cent of the world total; the key Middle Eastern countries are still relatively unexplored and have the potential for huge new discoveries, with Saudi Arabia alone claiming 150 billion barrels of 'yet-to-find'; and in 2005 OPEC produced almost 34 million barrels a day of the world's 81 million barrels a day.[3] But in all likelihood none of these statements is accurate, and all could be wildly misleading.

In the days when Western companies controlled the oil industry in OPEC countries, they would regularly report detailed reserves and production numbers in their annual accounts to satisfy their domestic financial regulators.[4] But after the nationalizations of the 1970s, although the oil kept flowing the information stopped. Since then most important data about OPEC countries' oil industries has been treated literally as a state secret. As a result even some of the most basic facts, such as how much oil OPEC actually produces from day to day, are almost impossible to divine. Members do declare their total production, but often months in arrears, and few in the oil market believe them. 'Everybody lies,' says Henry Groppe, the wise old Houston-based energy consultant and contemporary of Hubbert. 'The production information provided by the OPEC producers is absolutely unreliable.'

The reason is simple. One of the key criteria for deciding an individual country's quota within OPEC is its current level of production, and this gives all members an incentive to inflate the numbers they report. 'If you can get the others to believe your production is greater than it is, and if there's a bigger pie to be allocated, then you're going to get a bigger piece of it,' Groppe explains, 'and if OPEC as a whole needs to cut back, then you can just reduce paper production.'

The scale of the deception is extraordinary. For decades Groppe's firm has been laboriously reconstructing something closer to OPEC's real production numbers by collating oil import data from countries all around the world. These figures are much more reliable since duty is payable on imports, and it is in every importing country's financial interest to make sure they are accurately recorded. But collating the figures is a complicated business, and there are often long delays before governments publish the data, so the whole exercise can take as much as two years to complete: 'So about twenty-four months after the fact we know how much oil was imported in the world, and when we compare that with the oil that was reported as being exported, there's often as much as 2 million barrels per day of exports that never showed up any place. Well, obviously they were never produced.' In other words, as much as 2.5 per cent of total world oil production is simply a mirage, and every OPEC production number I have quoted so far – from the *BP Statistical Review*, which reproduces the OPEC figures – is probably far too high.[5]

If OPEC countries are ready to lie routinely about something that can – with a lot of effort – be checked, imagine how liberal their attitude might be towards figures that are effectively impossible to verify, should the need arise. On the face of it, this seems to be exactly what happened in the mid to late 1980s, when OPEC was discussing a proposal to change the criteria by which quotas were allocated to include the size of each country's reserves. The new rule was never actually introduced, but the prospect that it might be seems to have galvanized OPEC countries into suspiciously large revisions.

In 1985 Kuwait's proved reserves – the most stringent definition – leapt by almost half, from 64 gigabarrels (billion barrels) to 90Gb, and in 1988 they rose again to 92Gb. That same year Abu Dhabi's proved reserves almost *tripled* to 92Gb, matching Kuwait exactly, and then Iran raised the bidding by one, increasing its proved reserves from 49 to 93Gb, while Iraq more than doubled, from 47Gb to a nice round 100, and Venezuela also jumped by over 100 per cent from 25 to 56Gb. Finally in 1990 Saudi Arabia raised its proved reserves by a whopping 88Gb, from 170 to 258Gb. So in the space of five years OPEC reserves had risen by 305 billion barrels, despite the fact that no significant discoveries had been made. Most independent observers find this utterly incredible, not only because of the sheer enormity of the revisions, but also because of a string of other suspicious coincidences.

It was Dr Colin Campbell, the grand old man of peak oil, who first spotted them. He noticed that in 1984, just before the game of leapfrog started, Kuwait's declared reserves were 64Gb, and by that year it had produced 21.5Gb, meaning that the total discovered was 85.5Gb. The following year Kuwait increased its 'reserves' to 90Gb, and the closeness of the two figures led Campbell to suspect that Kuwait had simply started declaring the total oil it had ever discovered – including all the oil it had already produced – rather than its remaining reserves.

What was even more suspicious to Campbell was the fact that Kuwait, Abu Dhabi and Iran all declared nearly identical reserves, which he interprets purely as the result of quota competition. 'It is absolutely inconceivable that three separate countries should have exactly the same number! I think Kuwait is reporting the total that it ever discovered, but the others just picked a number out of the clear blue sky to be the same as Kuwait.'[6] More suspicious yet, many of the new reserve figures subsequently remained unchanged for many years, despite the fact that OPEC countries were producing billions of barrels every year.

Campbell and many other forecasters and industry databases now discount OPEC's reserve numbers by a substantial margin (table 5). To estimate something closer to their real reserves, Campbell starts with detailed industry data from about a decade ago, subtracts production since then, and adds any subsequent discoveries – although there haven't been many of those. So whereas Kuwait started with 64Gb before the revision race, and now claims 102Gb, Campbell credits it with just 54Gb. The difference is even starker than the bare numbers imply, since the OPEC figures are claimed as proved, whereas Campbell's are proved and probable, which ought to be larger.

None of this constituted proof of course, and for several years Campbell's view on the OPEC reserves, although persuasive, remained no more than conjecture. But then in January 2006 his analysis was vindicated from the most unlikely source – the state-owned Kuwait Oil Company. The authoritative industry newsletter *Petroleum Intelligence Weekly* got hold of some sensational internal company documents which revealed that Kuwait's 'remaining reserves' were less than *half* the publicly claimed figure, not 102Gb but 48Gb.[7] With that the world's reserves shrank by close on 5 per cent.

Table 5: How much oil does OPEC really have?

	OPEC	IHS ENERGY	CAMPBELL
Kuwait	102	52	51
Abu Dhabi	98	54	39
Iran	138	135	69
Iraq	115	99	61
Saudi Arabia	264	289	159
TOTAL	717	629	379

Table 5: Gulf OPEC members' claimed proved reserves for 2005 (*BP Statistical Review*, figures rounded) versus estimates of their proved and probable from IHS Energy, and Colin Campbell. The total difference today between OPEC and Campbell is 338 billion barrels, or more than 25% of the world's total reserves as reported by BP

The new number was even lower than Colin Campbell's estimate – although close enough to support his hunch – but even this wasn't the whole story. According to the documents, the result of an assessment in 2001 by the KOC's reserves management committee, the 48Gb covered both 'proven and non-proven reserves'. Kuwait's properly defined proved reserves were just 24Gb, less than a *quarter* of the publicly claimed figure. I used to feel that Colin Campbell was excessively cautious about OPEC's reserve figures, but in the case of Kuwait he turns out to have been too optimistic.

Strangely, however, the *BP Statistical Review*, the annual compendium of energy statistics, continues to carry the official figure for

Kuwaiti reserves of almost 102Gb, not the 24Gb that the Kuwait Oil Company evidently believes privately. BP ought to have a good idea what the real number is because of its long association with Kuwait; it held the concession until nationalization of the oil industry in 1975, and has had technical staff working with Kuwait Oil Company for the last fifteen years. But when I tried to ask Lord Browne which of the two numbers he actually believed, he smoothly evaded the question at two separate news conferences in 2006.

On the first occasion he passed the question to Tony Hayward, BP's head of exploration and production, who maintained there was a 'thin line' between resources and reserves, and that 'nothing we've seen would lead us to change our view that the resource base in Kuwait is robust and in line with what everyone was carrying previously.'[8] On the second he passed it to Peter Davies, BP's chief economist, who oversees the *BP Statistical Review*. When I pushed him on why they continued to publish the dubious official number, Davies averred, 'We always report the government numbers on everything, we do not second-guess anyone, and that is the objectivity of this review.'[9] It is a strange definition of objectivity to believe whatever you are told by somebody, even when they have a reason to exaggerate. But in the circumstances perhaps it would not be wise for BP to point this out: they lead one of three consortia vying to undertake Project Kuwait, the long delayed plan to ramp up the country's oil production, a contract that could be worth $7 billion.

The significance of the Kuwaiti revelation cannot be overstated. Not only are Kuwait's publicly claimed reserve figures shown to be highly questionable, but extreme doubt is cast on all the other OPEC reserve revisions of the 1980s. The difference between OPEC's claimed proved reserves and Colin Campbell's estimates of their proved and probable is 338 billion barrels of oil – more than a quarter of the world's reserves as reported in the *BP Statistical Review* – and the biggest single variable affecting forecasts of the global peak. If all OPEC's reserve numbers turn out to be as reliable as Kuwait's, any idea that the organization could supply the amounts of oil projected by the IEA, EIA, Exxon and Shell would be rendered absurd. It would also bring the date of the global peak significantly closer.

Within this massive discrepancy, Saudi Arabia alone accounts for more than a quarter of the disputed barrels. The Saudi increase is equivalent to 40 per cent of all the oil ever produced by the United States, where production began in 1860.[10] Oddly, however, I am less doubtful about the Saudi reserve revisions of the late 1980s than those of the other Gulf states. Although Colin Campbell is convinced that the Saudis are doing exactly what the Kuwaitis apparently did – presenting their total discovered as proved reserves – and therefore trims their figure from 264Gb to 159Gb, I still tend towards the Saudi number. I base this judgement largely on the testimony of two former senior Aramco executives, Sadad al-Huseini and his predecessor Edward Price, an American who oversaw much of the gradual transition to Saudi control. And both of them defend the Saudi reserve figures robustly. My reliance on them may sound naïve – *they would say that, wouldn't they?* – and perhaps it is. But both men are now retired, both have demonstrated their independence by criticizing other aspects of Saudi oil policy, and together they make a plausible case. Bear with me.

In Bahrain, al-Huseini described at some length the huge programme he instigated to assess the scale of Saudi reserves after the US oil companies left. At that stage he claims the reserves were 'grossly under-reported' because the Americans had never had any incentive to make a thorough assessment; it would have been expensive to carry out, and was unnecessary because at that stage reserves comfortably exceeded the needs of production. 'So they didn't spend much money or time worrying about ultimate reserves.' But when the Saudis finally took control of their oil resources they naturally wanted to know how much they were sitting on.

The new programme included investigative drilling to establish the full depth and width of the reservoirs, which had never been done before; testing rock samples from individual zones within each reservoir; extensive 3-D seismic surveys; and a massive computer modelling effort, 'the largest reservoir simulation programme in history, bar none'. All this work took several years, and it was for this reason that the hike in Saudi reserves lagged behind those of the other Gulf states. 'We were three years behind everybody, because we were doing it based on our studies, whereas many of our neighbours, Kuwait, Iran or Iraq, had done nothing virtually. All they did was take their reserves and multiply them by two. And they had done that ahead of us, so it sounded like we were imitating them.'

At his home near College Station, a couple of hours north of Houston, Edward Price took me through some of the detailed figures behind the Saudi reserve assessment, although of course I am not qualified to judge them. After a long career with Chevron, Price joined Aramco as Chief Petroleum Engineer in 1979, and rose to become Vice-President for exploration and production from 1984 to 1988 when – as Sadad al-Huseini's boss – he oversaw the Saudi reappraisal. 'We had all the data to justify it under the principles set up by the Society of Petroleum Engineers, their definition of reserves . . . The 258 billion figure we felt was proved without question, we knew we would get at least that.'[11] Price and his team were also confident the figure would grow to over 300 billion as they gained more experience of producing the fields.

Yet more controversy surrounds Saudi Arabia than any other OPEC member. Much of it has been generated by the Houston energy banker Matt Simmons, who is deeply sceptical about the future of Saudi oil production. His concerns are not principally about the size of Saudi reserves, rather the condition of its reservoirs, although of course the two are closely related. Simmons's doubts were first prompted during a visit to the kingdom in February 2003, just before the invasion of Iraq. He and a small group of Texas oilmen had been invited by Saudi oil minister Ali al-Naimi, who was evidently worried about deteriorating relations between his country and America following 9/11. During the trip the Texans were given a glowing presentation about Ghawar, the world's biggest oilfield, but for Simmons a couple of details jarred and piqued his curiosity.

There is no doubt that Ghawar is a truly astonishing field. Production there started in 1951 and since then it is estimated to have produced 55 billion barrels of oil and 5 billion barrels of condensate, more than the entire oil output of Europe to date.[12] Even now this single field is thought to produce about 5 million barrels a day, roughly half the total output of Saudi Arabia, and about 6 per cent of world production. On the map the field measures 30 miles wide by 170 miles north to south.

It was while Simmons was being shown a 3-D hologram of Ghawar that he noticed that all the production wells were clustered at the northern tip. When he asked why, a Saudi executive assured him that they were simply making an 'orderly march' from north to south. But later in the trip Simmons discovered that the reservoir quality – its porosity and permeability – declines markedly in the southern sections of the field, making him incredulous at the explanation he had been given. '*Orderly march*, give me a break!' he scoffs while recalling the event from his fiftieth-floor eyrie in downtown Houston. 'If you believe that,' he declares with a typical Simmons flourish, 'you better rent your tux for the wedding of the Easter bunny and Santa Claus.'[13] The reservoirs of Ghawar were not uniformly fabulous, he realized, and the Saudis had already had the best of them.

Simmons returned to Houston convinced there was something wrong with the Saudis' story, but feared that their secrecy meant he would never obtain the information to back his hunch. Then to his surprise he discovered that a series of technical papers about production problems in Saudi fields, written by Saudi Aramco geologists and engineers over several decades, had been sitting in the library of the Society of Petroleum Engineers in Dallas all along. As a banker, not a petroleum engineer, Simmons struggled to understand the papers, but eventually the result of his research was a book called *Twilight in the Desert*, which provoked a bitter row with Saudi Aramco.

Simmons's analysis runs like this: Saudi oil production has always been overwhelmingly concentrated in a very few giant oilfields, principally Ghawar; Saudi fields were probably 'over-produced' and damaged during various production spurts; the best production zones, such as the northern end of Ghawar, are substantially depleted, and the remaining zones and fields will be far less productive; and the successful use of techniques such as water injection and horizontal drilling has given an impression of invincibility, but has in fact accelerated depletion and increased the likelihood of an unexpected and rapid decline. As a result he concludes that overall Saudi production could soon peak – 'It would seem safe to conclude that Saudi Arabia's oil output is unlikely to grow in the coming years and could soon begin to decline' – or even, he suggests darkly, suddenly collapse.

Simmons's broader points are well made – although I am far less convinced by his 'sudden death' theory – and his withering attack has exposed some major weaknesses in the official Saudi position. News of the general thrust of his research led to a frosty confrontation in

February 2004 at a conference convened in Washington, where both he and Saudi Aramco officials made presentations to an expert audience. The Saudis – who evidently couldn't even bring themselves to address Simmons by name – were forced to release more data than ever before, although nothing like enough to douse the argument.[14] And during the course of the encounter, the Saudi officials made several crucial claims that now appear to be unravelling.

Dr Nansen Saleri, Saudi Aramco's manager of reservoir engineering, confidently declared that the company could produce 10, 12 or even 15 million barrels per day for the next fifty years without difficulty. To sustain the highest of the three projections they would need to replace 70 billion barrels of reserves, but this was 'very, very achievable'. The reason they were so confident of this, explained his colleague Mahmoud Abdul Baki, Vice-President for exploration, was that large areas of the country were relatively unexplored: 'We have a lot of acreage to explore, and the potential to find a lot more oil and gas.' In fact the company expected to raise its reserves by far more than necessary to support even its highest production scenario. Saleri said, 'We are looking very conservatively at upwards of 150 billion barrels over and above the 260 billion barrels that we carry as proved reserves right now.'

The astonishing claim that Saudi Arabia's proved reserves will 'soon' rise by another 60 per cent has since been repeated in speeches by oil minister Ali al-Naimi, but the Saudi officials provided no justification for it in Washington, and the figure is wildly higher than any other estimate. The United States Geological Survey's *World Petroleum Assessment 2000* estimated that Saudi Arabia had 87 billion barrels of 'undiscovered resources', but even this number is likely to be substantially over-inflated, since the global projections contained in this report have already proved far too optimistic, and its methodology has been widely criticized.[15] Edward Price prefers to cite an exhaustive but unpublished study carried out for the Saudis by the four American companies that originally made up Aramco (Exxon, Chevron, Texaco and Mobil) which assessed the entire country in the mid-1980s. On the basis of intensive field work it concluded there were just 16 billion barrels left to find. So the man who was once in charge of exploration and production for Aramco has no confidence in the USGS estimate, and still less in the Saudis' new figure, which he judges 'highly improbable' and inflated to scotch Matt Simmons's attack.

Price also debunks the Saudi claim that large areas of the country are substantially unexplored. 'We shot seismic across pretty much the whole country looking for big fields, but most of the structures we found had no oil. There may be a bunch of little bitty fields up towards Kuwait, but they won't amount to much.' This assessment seems to be supported by what is known of the recent exploration history of Saudi Arabia, as Matt Simmons reports: the Saudis keep announcing new finds, although they never reveal their size, and the last supergiant discovery – Shaybah – was in 1968. It would also fit with what is known of the rather depressing exploration trends in OPEC as a whole: over 1,000 fields have been found in OPEC countries since 1980, of which only 10 per cent were larger than 130 million barrels and 50 per cent were smaller than 8 million barrels.[16] All of which makes the idea of 150 *billion* new barrels in Saudi Arabia seem even more far-fetched.

Clearly the fewer of those new barrels that are forthcoming, the less likely it is that Saudi Arabia can maintain oil production at the levels and for the duration that it claims. What's more, the higher Saudi Arabia tries to push its production rate, the greater the geological and engineering challenges will become. Saudi Aramco estimates its existing production has a 'natural decline rate' of 6 per cent, meaning that if – hypothetically – the company stopped all maintenance and drilled no new wells, its production capacity would fall by 6 per cent over the course of a year.[17] That means that if the company produces oil at a rate of 10 million barrels per day, it has to add an extra 600,000 barrels of daily capacity each year just to maintain that rate. But if it somehow managed to push production up to 15 million barrels per day, it would have to add 900,000 barrels of daily capacity each year just to stand still. And that's the equivalent of building the entire oil production industry of Angola or Qatar every twelve months.

Saudi Arabia claims it will increase its production capacity from 11 million barrels a day in 2006 to 12.5 million barrels a day in 2009, but many doubt it can achieve this on engineering grounds alone. Sadad al-Huseini told Reuters in November 2005 that the country 'may wind up two to three years behind schedule' simply because of the shortage of rigs and equipment. He went on to point out even more fundamental difficulties: 'Every new increment is going to be more expensive and complex and yield smaller amounts. And you will have to replace easy production and large capacity declines with much more difficult production.' He also disparaged the production forecasts such as the IEA's which suggest that Saudi production could rise to almost 18 million barrels per day: 'It is not clear why in the next 20 years Saudi Arabia would want to go above 13.5 million bpd with all its technical risks and consequences. People who say otherwise are using very simplistic assumptions and are not talking about how production can be sustained.'[18] Edward Price goes further, arguing that Saudi Arabia is unlikely to be able to maintain output for anything like as long as it claims, even at lower levels. 'My view is that they shouldn't try to go any higher than 12 million barrels a day,' he told me in Texas. 'They could probably get to 15 but it would be a massive stress on the system and they couldn't hold it very long. Even at 12 they'll only hold it for ten or fifteen years before it goes into decline.'

Al-Huseini's temerity provoked a haughty rebuke from the Saudi state. A statement from the Ministry of Petroleum, reported by the *International Oil Daily*, claimed his remarks were 'false and baseless' and motivated by 'personal goals' (unspecified), and concluded: 'His statements are not only misleading and wrong, but they also cast doubts on the capabilities of the Saudi petroleum industry, and they constitute an insult to the kingdom's inalienable policy and role for realizing the world petroleum market.'[19] Now then, who sounds rational and who sounds rattled?

It is not only former employees who question Saudi Arabia's oil policy. Even current officials and executives – who presumably face even greater risk of punishment for speaking out – are beginning to question the kingdom's self-proclaimed production targets. On the very morning of the Saudi-Simmons confrontation in Washington, the *New York Times* ran a piece under the headline 'Forecast of Rising Demand Challenges Tired Saudi Fields', in which anonymous Saudi officials disparaged the official optimism: ' "We don't see us as the ones making sure the oil is there for the rest of the world," one senior executive said in an interview. A Saudi Aramco official cautioned that even the attempt to get up to 12 million barrels a day would "wreak havoc within a decade", by causing damage to the oilfields.'[20]

More recently the *Financial Times* reported that Saudi officials have warned that OPEC will fail to meet the demand growth predicted by Western agencies within ten to fifteen years. The story explained that the International Energy Agency's forecasts depended on the assumption that OPEC could raise its production to 50 million barrels per day by 2020, but that its officials had been told this was highly unlikely:

> senior Saudi energy officials have privately warned US and European counterparts that OPEC would have an 'extremely difficult time' meeting that demand. Saudi Arabia calculates there is a 4.5m b/d gap between what the world needs and what the kingdom can provide . . . Saudi Arabia pumps 9.5m b/d and has assured consumer countries that it could reach 12.5m b/d in 2009 and probably 15m b/d eventually. But a senior western energy official said: 'They said it would be extremely difficult to move above that figure.'[21]

Distrust of the new Saudi numbers, and of the OPEC revisions of the 1980s, is now evident even in the official publications of the IEA. Because it represents the OECD oil importers on energy matters, the Agency has to maintain the diplomatic niceties, yet its disbelief is now palpable. Its *World Energy Outlook 2005: Middle East and North Africa Insights* pointedly cites not only OPEC's claimed reserve figures, but also the much lower numbers from IHS Energy. And the report is peppered with phrases such as 'if reserves are indeed as large as official estimates show' and, referring to the Saudi oil minister Ali al-Naimi, 'these projects, he has claimed, would allow production of 15 mb/d to be sustained for at least 50 years'. In other words, even the IEA now seems to be reluctant to take his word for it.

It's not hard to see why Saudi Arabia might want to exaggerate the scale of its oil resources; the position of its royal family demands it. For decades the fundamental contradiction of Saudi Arabian society has been that a population that largely adheres to Wahabism, the most puritanical, fundamentalist and xenophobic sect of Islam, is ruled by a royal family that is venal, bloated and Westward-leaning. The absolute monarchy has only managed to survive because its control of the oil has allowed it to buy American protection, and to bribe its own people. But now the entire arrangement is teetering.

With the arrival of the oil boom of the 1970s Saudi Arabia's royal family was able to buy domestic political legitimacy, or at least acquiescence, by providing one of the most lavish welfare systems anywhere in the world. For years Saudis effectively paid no tax, and received not just free healthcare and education, but also subsidized housing, electricity, fuel and food, and interest-free loans – often never repaid – for anything from property speculation to ludicrous schemes to grow wheat in the desert.[22] But since their heyday in the early 1980s Saudi finances have deteriorated badly, and social benefits have been cut back. And today – paradoxically at a time of record high nominal oil prices – the Saudi state is less capable of buying off its people than ever.

Saudi Arabia's revenue from oil exports, which account for three-quarters of its income, collapsed from $22,600 per capita in 1980 to just $4,600 in 2004, largely because its population tripled to almost 22 million over the same period.[23] According to the CIA half the population is under twenty-one, and unemployment is thought to be as high as 25 per cent. There are frequent acts of terrorism and attacks on the security forces, but opposition is not restricted to Wahabi fundamentalists; it is also growing among the new, Western-educated middle class.[24] The house of Saud has every reason to be terrified that it will soon go the way of the Shah of Iran. In these circumstances the royal family needs American protection more than ever before.

Saudi Arabia's security pact with America was famously sealed when Roosevelt met Ibn Saud on his way back from the Yalta conference in 1945. Saudi oil production was negligible at the time, but US officials had already concluded that Saudi reserves constituted 'a stupendous source of strategic power, and one of the greatest material prizes in human history'.[25] Although the meeting on board a US warship in the Suez Canal went unminuted, it is widely thought the two reached an understanding that has apparently survived to this day: America would guarantee the security of Saudi Arabia – and in effect the house of Saud – and in return the Saudis would guarantee US access to their vast reserves. In any event, successive US presidents have behaved as if that was the arrangement, providing the Saudis with billions of dollars in arms for both national defence and internal security. And Bush the Elder took America into the first Gulf War explicitly on that basis, declaring 'the sovereign independence of Saudi Arabia is of vital interest to the United States'.[26]

Over the years the US has put up with all sorts of humiliations and setbacks for the sake of the deal: the Arab oil embargo; the nationalization of Aramco; the embarrassment of Saudi Arabia's record of tyranny and torture; and even, it seems, the revelation that most of the 9/11 hijackers were Saudi. America swallowed them all because, except for 1973, the house of Saud always kept the oil flowing. But what possible use would the US have for a regime that couldn't or wouldn't? The house of Saud has every reason to make the Americans think it can raise production, even if that increase is short-lived, or in fact impossible.

Chris Skrebowski is now the editor of the *Petroleum Review,* but he previously worked for eight years as an oil analyst for the Saudi Ministry of Petroleum, and has a keen understanding of the political context in which the Saudi oil industry operates. He is convinced that it is the vulnerability of the Saudi royal family that leads it to make claims that horrify the country's oil industry professionals. And this is the reason for diametrically opposing stories coming out of Saudi Arabia.

> The technocrats know perfectly well these higher production numbers aren't possible and they're getting ever more nervous because they know that if they're called on to even try they will start really damaging the fields. The royals don't care because they're staring extinction in the face and are more or less prepared to promise anything to anyone. They may or may not know that it's lies, but if it gives them a few more years, they'll tell the Americans anything they want to hear.[27]

Saudi-US dependency appeared to be solidly mutual until early 2006. But then President Bush apparently renounced decades of American energy and foreign policy in a jaw-dropping State of the Union address. 'Keeping America competitive requires affordable energy,' he declared. 'And here we have a serious problem: America is addicted to oil, which is often imported from unstable parts of the world.' Setting a target that America should replace more than 75 per cent of its oil imports from the Middle East by 2025, Bush claimed the country could 'move beyond a petroleum-based economy, and make our dependence on Middle Eastern oil a thing of the past'.[28] For a Republican President and former oilman it certainly sounded like a radical departure, the idea of energy independence plays well to American insecurities, and it won Bush a standing ovation or two. But this apparent policy reversal as presented has so little chance of success that it must suggest a deeper agenda at work.

America guzzles oil so extravagantly that of course it could cut back significantly if there was ever the political will. In round numbers America consumes 21 million barrels of oil per day, importing a total of 13, of which the Persian Gulf supplies 2.5.[29] Of the 21 million barrels per day, more than 9 million are consumed as petrol, and the average American car does just twenty-four miles per gallon. So by doubling the average fuel efficiency of its light vehicle fleet – if everyone drove a Toyota Prius, or even my old Peugeot diesel estate – the US could in theory save 4.5 million barrels per day, almost double its imports from the Middle East. But of course that would mean depriving voters of their SUVs, and no American politician could do that and hope to get re-elected.

Yet having set an import reduction target, the Bush administration remained obstinately opposed to the very measures that could achieve it, such as legislating to raise CAFE (Corporate Average Fuel Economy) standards, or imposing higher gasoline taxes. There was no hint of measures to control oil demand in the speech, and the President baldly asserted that the best way to break America's addiction was through 'technology'. But the areas he picked out for additional research funding were ethanol and hydrogen which, as we saw in chapter 4, are not remotely capable of moving America 'beyond a petroleum economy'. In a coruscating editorial the *New York Times* damned the measures as 'woefully insufficient', and the *Financial Times* called them 'wishful thinking'.[30]

Although the announcement was couched in terms of 'Middle East oil', it was especially hostile towards America's longstanding ally Saudi Arabia, since it supplies two-thirds of American imports from the region. The message was sharpened by the fact that the speech was delivered on the very day that Saudi Arabia blocked a proposal from Iran and Venezuela to cut OPEC production. The Saudis were clearly shocked, and must have thought that their sixty-year-old oil-for-security pact was beginning to unravel. And by raising the possibility of lower demand for Middle Eastern oil, the administration had given Saudi Arabia and the other producers reason to hold back from expanding production capacity – even assuming they are able to – which could bring about a shortage long before the US could ever hope to have trimmed its consumption. The policy was not only toothless but might also be counterproductive. So what did they hope to achieve by it?

The speech starts to make more sense in the light of an apparent evolution in the administration's thinking on energy which took place the previous year. During the summer of 2005 the administration made great efforts to canvas the views of major oil companies and energy forecasting consultancies about the outlook for worldwide oil production. The process culminated in a series of intimate meetings at the White House, where one by one the companies briefed Dick Cheney on their analysis. Some of those who presented to the Vice-President got the impression he was already very well informed on the subject. 'Dick Cheney asks great questions,' recalls Mike Rodgers of PFC Energy. 'They're very specific: *What do you think about this?* or *What's your number on that?* It was obvious that he had heard a lot of this stuff before.'

The major oil companies refused to divulge any details of their involvement in the consultation, but evidently a common theme among several participants was the likelihood that non-OPEC production would plateau after 2010, and that from then on all global demand growth would have to be satisfied by OPEC. Rodgers and PFC chief executive Robin West went on to raise with Cheney the question of whether or not OPEC could actually deliver all that additional oil. 'Our message was we're not here to tell you one way or another that OPEC *can't* do it, but we're sceptical,' says Rodgers. 'We wanted to get the point across that OPEC is going to have to come up with a lot more production in 2015 and beyond. And if they can't and we don't prepare for that possibility then you've got a real problem.'

Rodgers and West pointed out that it is inherently difficult to forecast OPEC production because its reserve figures are opaque and potentially untrustworthy. What they could say with some certainty, however, was that OPEC is producing 8 billion barrels a year more than it finds. Given this massively negative balance, and given that PFC's analysis of other oil-producing regions confirms that production usually starts to fall at about the halfway point, there was good reason to doubt OPEC's ability to meet future global demand. Indeed, taking a more conservative view of the organization's likely reserves, OPEC could reach 50 per cent depletion within ten years. In other words, PFC Energy warned Dick Cheney that even though everything now relies on OPEC, its production might peak within a decade.

Cheney didn't give much away during the course of their meeting, West told me: 'He just rubbed his chin and listened, which is Dick's usual way.'[31] If the administration has taken this idea on board, however, the State of the Union reversal begins to make rather more sense, although the policy measures themselves remain just as ineffectual. In this light the declaration looks more like an attempt to soften up the American public to the idea of an impending oil shortage, while casting the Middle East producers as the villains of the piece. No policies that hurt just yet, but lay down a political marker, get people used to the prospect, and start spinning who's to blame when the crisis finally begins to bite.

The date at which OPEC production will actually start to decline is difficult to forecast with any confidence: some OPEC members lie about their production, some OPEC members lie about their reserves, and some OPEC members evidently pluck yet-to-find and production forecasts out of thin air when it suits their purpose. But the very fact that they all seem to be dissembling must increase the risk that the OPEC peak will come sooner rather than later. So it seems that, contrary to popular opinion, President Bush has every chance of hitting his target. If the doubts of PFC Energy and others are right, it's likely that America – and the rest of the world – *will* soon begin to find their Middle East oil imports falling. It just won't be voluntary.

8

Interesting Times

THE LAST OIL shock might easily have started on 24 February
2006 – regardless of when the peak in production actually tran-
spires – and the fact that it didn't may owe something to the sacrificed
lives of two security guards in Saudi Arabia. Suicide bombers drove
three cars packed with explosives to Abqaiq, the world's largest oil
processing plant, and rammed the gates. Officials insist that the attack-
ers were stopped at the perimeter checkpoint of the facility, a mile
from the plant itself, but relatives of the guards said the militants
managed to get through to a second checkpoint. In any event, the cars
exploded in the vicinity of a plant that processes 6 million barrels a
day – more than three times the UK's total oil output – and the attack-
ers were killed only after a two-hour gun battle, during which the two
guards also died. A statement purporting to be from al-Qaeda in the
Arabian Peninsula claimed responsibility and promised to keep trying:
'We renew our vow to crush the forces of the crusaders and the tyrants
and to stop the theft of the wealth of the Muslims.'[1]

I use the term *last oil shock* to describe not the peak of global oil production but the all-encompassing crisis it is almost bound to precipitate. This crisis will be financial, economic and geopolitical, and need
not coincide with the date of the peak itself, which is discussed in the
next chapter. It could break long before or even some time afterwards.
As the hurricanes showed in 2005, when there is very little spare oil
production capacity, relatively minor interruptions to supply can
become a big problem. Katrina and Rita gave us just an inkling. Had
al-Qaeda succeeded at Abqaiq, the supply disruption and resulting
financial chaos could have caused a deep global recession all by itself.
That would have deferred the oil peak by depressing consumption
possibly for some years, but the last oil shock would already have started.

Whatever the precipitating event turns out to be, what will almost certainly follow is a period of extreme oil price volatility, as demand repeatedly hits the ceiling of production capacity, whether that is determined by short-term events above ground or the fundamental geology. This is likely to set off recurrent global economic slow-downs or recessions, which in turn could have the effect of smearing out what would otherwise be a relatively sharp oil production peak into a more extended and corrugated plateau. The geological peak may well be delayed for a time, but given the cause this will be no consolation. And when it does take place, the passing of the peak may well be obscured by the static and interference of short-term events, until the oil production decline is finally and definitively established. As the Houston energy banker Matt Simmons is fond of saying, we will only see the peak in the rear-view mirror. Whether we could afford to run the car at all by that stage is another question.

Paradoxically the very worst short-term outcome might be not a sudden shock, but a milder recession. If this were to create some temporary spare oil production capacity by depressing demand, the economists would claim it was all back to business as usual, and the urgency of the need to prepare for the impending peak could easily be forced off the policy agenda – assuming it ever gets there – by apparently more pressing problems. The world might roll over and go back to sleep again, only to suffer a far more brutal awakening later on. But somehow I doubt we will have the luxury of the lie-in.

As the global peak approaches and the market tightens, any sudden interruption of oil production could spark the last oil shock. There are plenty of potential flashpoints: an upsurge in rebel activity in Nigeria, which has already cut the country's production by as much as 750,000 barrels per day, or more than a quarter; the shutting-in of another major field such as BP's Prudhoe Bay, closed in 2006 with the loss of 400,000 barrels per day because the company had allowed its pipelines to rust to the point of failure; or simply another bad hurricane season in the Gulf of Mexico. The biggest single threat of course stems from the unfolding confrontation between the US and Iran, which could still engulf the entire Middle East, despite a relative softening of the rhetoric more recently.

On the face of it the fight seems asymmetrical. In April 2006 Seymour Hersh, the respected American journalist who broke the Abu Ghraib story, reported that the White House was preparing to attack the Islamic Republic to effect regime change, regardless of the outcome of the nuclear 'diplomacy', and was planning to do so with tactical nuclear weapons.[2] Yet Iran is anything but defenceless. Only a week before the Hersh story was published, the oil price had jumped by $2 a barrel after the Iranian navy test-fired a newly developed torpedo during manoeuvres in the Strait of Hormuz, the narrow waterway at the mouth of the Persian Gulf through which a fifth of the world's oil supplies pass every day.[3] The demonstration was clearly intended to reinforce the Islamic Republic's earlier threat to respond 'by all means' and with 'severe consequences' to any military attack by the US or Israel.[4]

Iran's 'means' include not only disruption to shipping in the Persian Gulf, and perhaps attacks on the oil production facilities of its neighbours, but almost certainly a reprise of Tehran's most powerful tactic, the withdrawal of its own oil exports from the world market. The last time Iran did this, during the revolution in 1979, the price of crude more than doubled.

Iranian exports supplied a much bigger proportion of the world oil market back then, but against that the market now has far less ability to absorb any loss. During the second oil shock when Iran cut off exports of about 4 million barrels a day, Saudi Arabia was able to ramp up its production almost instantaneously by 2 million barrels a day, and other OPEC members also increased their output.[5] Today, however, there is very little spare capacity in OPEC or anywhere else. So although Iran's current exports of about 2.7 million barrels a day are lower than in 1979, their withdrawal could be just as damaging, if not more so. If the oil price were to double again, it would start from a much higher base.

In fact the US-Iran crisis is likely to be far more dangerous than 1979; during the second oil shock Iran was completely isolated, even within OPEC, but since then it has cultivated some seriously powerful friends, such as China. The world's second biggest oil consumer already buys 13 per cent of its supplies from OPEC's second largest producer, Chinese companies have won major construction projects in Iran, and diplomatic relations are warm. In 2004 the bond grew even stronger when the two countries signed a $70 billion memorandum of understanding under which China will help develop Iran's massive Yadavaran field, and buy 250 million tonnes of Iranian LNG over the next twenty-five years. There are also plans to build an oil pipeline from Iran to China through the Caspian and Kazakhstan. Iran is a critically important source of hydrocarbons to fuel China's relentless economic growth, so by picking a fight with the Islamic Republic, America also challenges its superpower ally.

This dynamic is unlikely to weaken, despite the apparent retreat of the Neocons in America. Whatever its precise political complexion, Washington will almost certainly continue to see China as a threat to US energy security, and has even set up a permanent body to monitor the issue. In November 2005 the annual report to Congress of the US–China Economic and Security Review Commission warned of the challenge to American interests posed by China's rapid military modernization, astonishing economic growth and its energy policies. The report noted that China was now the second largest oil consumer after the US, that some 40 per cent of global oil demand growth now comes from China, and that Chinese oil demand was expected to keep rising by almost 6 per cent a year for the next decade. This meant that China was 'on course to compete with the United States and other oil importing nations for global supplies'.[6]

In particular the Commission attacked China's 'mercantilist' approach to securing oil supplies, meaning that instead of buying barrels on the open market, China buys up entire oil and gas fields around the world exclusively for its own use. The report noted with some alarm that China has done these kinds of deals not only in Iran, but also in Australia, Azerbaijan, Burma, Ecuador, Indonesia, Iraq, Kazakhstan, Oman, Peru, Sudan, Yemen, and even in America's own back yard, in Venezuela and Canada. One witness to the Commission testified that 'Every barrel of oil that China buys in America, whether it is in North America, Central America or Latin America, essentially means one less barrel available for the US market.' And the Commission's chairman Richard D'Amato told Congress: 'It is critical to persuade China to abandon this mercantilist spree to lock up attractive energy supplies wherever it can.'[7] While the Commission tepidly recommended a co-operative approach to reduce both countries' reliance on imported oil, it nevertheless urged that America should respond 'as aggressively as necessary to protect important U.S. interests'. As if the Bush administration needed any such encouragement.

The US attack on Chinese mercantilism is more than somewhat ironic, since America behaves in exactly the same way when it perceives its interests are threatened. When a bidding war broke out in 2005 between Chevron and the Chinese company CNOOC for control of Unocal, a mid-sized American independent, the issue was settled not by shareholders in a free market, but in Washington. The reaction in Congress was so hysterically nationalistic that CNOOC was forced to withdraw, leaving Unocal to fall to Chevron – for less money than the Chinese had offered. We already know that American policy is driven by an awareness of the impending global peak, and China's actions suggests that it too understands the situation perfectly well. Both sides behave as if an oil shortage is looming, and that 'it's us or them', although their approaches differ. While the US rails at China's mercantilism, I haven't noticed the People's Republic invading any country to liberate its oil resources. At least, not yet.

America is most threatened by China's inroads into the Middle East, where most of the rest of the oil lies. The US-China Economic and Security Commission report noted that Saudi Arabia is now China's largest oil supplier, that Saudi Aramco owns 25 per cent of China's biggest refinery and petrochemical project, and that China had been given rights to explore for gas in the kingdom. 'The United States is heavily reliant on Saudi oil and gas, and therefore Saudi sentiments toward our nation are of considerable import. While there is still a strong US-Saudi relationship, Beijing would like to see Riyadh shift its friendship to China. It is important for the United States to be mindful of this and to ensure that China does not undermine the US-Saudi relationship.'[8]

Worse still from America's viewpoint, while China 'locks up' increasing amounts of oil and gas from Iran – the country with the world's second largest reserves of both, at least according to official figures – US companies are excluded from doing business in the Islamic Republic by the Iran-Libya Sanctions Act (ILSA). So Iran presents the Bush administration with the same sort of quandary it faced over Iraq, only more so; Iran's energy deals with China are far more extensive and advanced than Iraq's with France or Russia before it was invaded. The American academic Professor Michael Klare, author of *Blood and Oil: How America's Thirst for Petrol is Killing Us*, concludes: 'From the Bush administration's point of view, there is only one obvious and immediate way to alter this unappetizing landscape – by inducing 'regime change' in Iran and replacing the existing leadership with one far friendlier to US strategic interests.'[9] If you doubt this analysis, it is worth remembering how Dick Cheney railed against the effects of US-imposed sanctions on the oil industry (chapter 1), and that of the three major oil-producing countries from which American companies were excluded at the time the Bush administration took power, only Iran remains legally closed to US interests. Not for long, I suspect.

After the catastrophe of Iraq, however, you might think that even the Bush administration couldn't be so stupid, or that, following the Republicans' drubbing in the mid-term congressional elections in November 2006, it wouldn't dare. But although the likelihood of such an attack does seem to have receded, at least for the time being, it would be easy to overestimate the importance of this shift. Regardless of which party controls Congress or forms the next US administration, the causes of this potentially epic conflict will inevitably sharpen. Competition between the US and China for the remaining supplies in a handful of Middle Eastern countries can only intensify as oil production peaks, first in the non-OPEC world and then globally. At some point, therefore, China may well be forced to decide how to react to an attack on a key ally and energy supplier, and this has the makings of another Cuban missile crisis. The great irony is that America, in its desperation to achieve 'energy security', will have prematurely destroyed both its own and everybody else's, and set off the last oil shock.

Whether the last oil shock is sparked by the US-China confrontation over Iran, or by some other geopolitical spasm, its effects are likely to be swift and brutal. The fundamental effects of the oil peak have already been explored in chapter 5, but long before they assert themselves, the first major oil price spike is likely to send a series of violent quakes through the economy. Quite how violent will depend on the level of awareness of investors on the currency and stock markets. For as long as most continue to labour under the misapprehension that every short-term spasm in the oil supply is simply some disconnected local difficulty, the response of the financial markets may be relatively subdued – perhaps with the exception of Iran. But the moment the money men *get it*, the price of oil and other energy assets will soar, and almost everything else will go into meltdown.

A major spike in the oil price is recessionary not only because of its direct effects on the global economy, but also because it is likely to cause stock markets around the world to crash, further reinforcing the recessionary pressures. This in turn will lead to second order effects, such as the deepening insolvency of many pension funds, which hold the bulk of their investments in stocks and shares. The fallout is likely to be particularly bloody in Britain, where the Pensions Regulator is monitoring 300 major occupational schemes that are already in danger of going bust.[10] As the crisis deepens, pension payments may be slashed to derisory levels in both money purchase and the supposedly more secure final salary schemes. The value of endowment policies will collapse too, with devastating effect on the borrowers who were counting on them to repay their mortgage, and the housing market as a whole. The banking sector will also act as a multiplier: since so much lending is 'secured' against future economic growth, as the outlook worsens lending will fall, leading to further contraction.

The potential economic impact of the last oil shock is analysed in some detail in a report commissioned by the US Department of Energy called *Peaking of World Oil Production: Impacts, Mitigation, and Risk Management*. The study made no prediction about the date of the peak itself, but set out to calculate how long it would take to implement measures to mitigate its worst effects, on the assumption of a massive, government-directed crash programme such as the Manhattan Project. Its overall conclusion was stark: 'The world has never faced a problem like this. Without massive mitigation more than a decade before the fact, the problem will be pervasive and will not be temporary. Previous energy transitions (wood to coal, coal to oil) were gradual and evolutionary; oil peaking will be abrupt and revolutionary.'[11]

'When we handed it in, the reaction was, "Oh my God",' recalls the report's principal author Robert Hirsch at his home in Washington DC in September 2005. 'It laid out the absolute enormity of the problem, and they didn't know what to do with it. People were truly frightened.' So much so that it took six months for the document to be cleared for publication, and there is still no sign that its recommendations are being acted upon.

Hirsch, a silver-haired engineer and physicist who spent most of his career managing research and development programmes in both government and the oil industry, conducted the study with two energy economists – 'they're pragmatic, realistic, not religious economists,' he says laughingly – and still the conclusions were dire. 'It could easily be the worst thing that's happened in our lifetimes. You're talking about a major catastrophe, something that looks like the Great Depression.'

In one respect, however, Hirsch expects the last oil shock to look less like the 1930s and more like another dismal decade, since he predicts the soaring oil price will produce not only recession but also high inflation, the rare economic condition known as 'stagflation'. The report drew widely on the experience and analysis of the first two oil shocks: 'As peaking is approached, devising appropriate offsetting fiscal, monetary, and energy policies will become more difficult. Economically the decade following peaking may resemble the 1970s, only worse, with dramatic increase in inflation, long-term recession, high unemployment and declining living standards.'

The most dangerous agent of economic contagion could be the US dollar, which is already sliding in value and widely thought to be on the verge of collapse. James Turk, a former banker with Chase Manhattan and investment manager of Abu Dhabi's oil wealth, marshals the arguments convincingly in his recent book *The Coming Collapse of the Dollar and How to Profit From It*. Turk argues that all fiat – or government controlled – currencies are prone to being debauched precisely because they are controlled by governments. Trapped between the competing and contradictory demands of taxpayers on the one hand and the beneficiaries of state spending on the other, all governments tend to take the easy way out and borrow irresponsibly. This leads to rampant inflation, the collapse of currencies, recessions and worse. The fall of ancient Rome, the end of the Weimar

Republic, and the collapse of the Argentine economy are all good examples of the destructive power of currencies subverted by spend-thrift governments. None of them, however, was quite as profligate as the United States today.

The dollar is now groaning under unprecedented and astonishing levels of debt, run up by all sections of American society. One big culprit is big government: more Americans are now employed by the government than in manufacturing – although this also reflects the fact that US industry has shrunk massively as companies have shifted factories overseas to exploit cheap labour.[12] Nevertheless, true to Turk's thesis, total government borrowing has soared from about $500 billion in 1970 to over $8 trillion in 2005. American citizens have also been on a two-decade spree fuelled by credit cards and equity withdrawal and they too are now neck-deep in personal debt. US companies have also borrowed massively, and by 2003 their exposure to risky financial instruments known as derivatives stood at an eye-watering $100 trillion. Warren Buffett, the legendary but normally mild-mannered fund manager known to investors as the Sage of Omaha, has said that derivatives pose a 'mega-catastrophic risk' to the financial system.

By 2005 American society as a whole owed around $39 trillion, three times the size of US GDP, or $135,000 for every man, woman, child and infant.[13] But although American profligacy is unique, Turk argues, Europe and Japan are also living way beyond their means. 'The stage is set', he concludes, 'for a currency collapse à la Weimar Germany or 1990s Argentina . . .'

A dollar collapse would devastate the US economy by making imported goods much more expensive, and forcing over-extended US consumers to retrench. This in turn would export recession around the world by slashing American demand for imported products. Rising US import prices could also increase inflationary pressures and lead policymakers to raise interest rates, so exacerbating the downturn.

Turk's analysis is widely echoed around the financial markets – the former chairman of the Federal Reserve Paul Volcker has also warned that the current position is unsustainable – and even among international authorities. Although it doesn't use quite the same language, the International Monetary Fund (IMF) appears to agree. Its *World*

Economic Outlook report for 2006 concluded that high oil prices were worsening economic imbalances around the world and 'heightening the risk of a sudden disorderly adjustment' – diplomatic code for a dollar collapse. And that, said the IMF's managing director Rodrigo Rato, would be 'very costly and disruptive to the world economy' – official-speak for a deep global recession.[14]

All this may sound alarming enough – and so it should – but for the United States there is even more at stake: nothing less than the survival of its hegemony. Ever since the early 1970s, when Henry Kissinger cooked up a deal with the Saudi royal family, oil has been bought and sold internationally almost exclusively in US dollars, and this arrangement has brought America immeasurable advantage. Oil-producing nations that receive dollars for their crude have to find somewhere to put their surplus greenbacks, and invariably this means spending them in the United States. Either they lend them back to the US government by buying US Treasury bills – helpfully keeping American interest rates lower than they would otherwise be – or they purchase goods made by American companies, vast quantities of arms being a notable favourite. The process is known as 'petrodollar recycling', and it allows the US to print dollars like it's going out of fashion, and gives the country an almighty free ride in both oil and debt markets.

If America had to pay for its oil in a foreign currency, as every other country does, it would have to sell dollars to buy that currency. And having used the currency to buy the oil, it might well never see the money again. But since it pays in dollar bills, which come winging back like homing pigeons, America effectively gets something for nothing; it can have its dollars *and* spend them.

If America benefits because it pays for its oil imports with its own currency, it benefits again because everybody else is forced to pay in dollars too. Since most countries import oil, their central banks have to maintain a stash of dollars to cover the cost. Again, the quantities are so large that many of these dollars have to be reinvested in Treasury bills, so further depressing US interest rates and allowing America's government, consumers and companies to continue their borrowing binge.

This effectively 'captive' financing is also the only thing that props up a yawning US trade deficit, which otherwise would already have brought the dollar crashing down. The US is only able to manufacture fewer and fewer of its own goods, and to import and consume vastly more than it exports, because other countries that actually make things lend it the dollars it needs to cover the difference. In the decade to 2005 the annual US trade deficit rose from about $100 billion to $724 billion, taking its total net foreign debt to $5.25 trillion, some 40 per cent of US GDP. It was only able to do so without the dollar collapsing because central banks around the world massively increased their dollar reserves. China's stash, for instance, soared from negligible to almost $1 trillion by September 2006.[15] As one astute commentator put it: 'in a nutshell, the US produces dollars, while the rest of the world produces things that dollars can buy.'[16]

The fact that the world's central banks hold so many dollars of course gives them a distinct interest in maintaining the value of the currency. As worried as they are about their level of exposure, if they tried to bail out quickly they would inevitably suffer huge losses as a plunging dollar slashed the value of their remaining reserves. As a result the bubble continues to inflate, and in the absence of a pin some analysts judge that it could continue to do so for some time. Chris Sanders, a Texan hedge-fund manager based in London, and the founder of Sanders Research Associates, concludes: 'It is not obvious that the dollar will collapse imminently, but if it did collapse it would be easy to see why. If these numbers continue to grow in this direction something very drastic is going to have to happen.'[17]

The dollar has already begun to stagger under the weight of American debt, falling almost 20 per cent between 2002 and 2005.[18] Since then oil producers have begun to talk of ditching the US currency, and both producers and central banks have quietly started to diversify out of it. In April 2002 OPEC's then head of petroleum market analysis, Javad Yarjani, gave a speech in Spain setting out the conditions that would be required for OPEC to adopt the euro as its oil trading currency.[19] The following year Indonesia's state oil company Pertamina said it might switch to euros, and in June 2006 rouble trading of oil futures started on the Moscow stock exchange. More recently the dollar has lurched on any news that central banks or oil producers were cutting – or even thinking of cutting – their dollar exposure. So far Sweden has halved its dollar reserves, Qatar has announced that it will raise the proportion it holds in euros, and Kuwait and the UAE are thinking about it.[20] If China did the same the shift could be seismic.

The financial and economic impacts of the last oil shock will of course be suffered globally, but the United States is particularly exposed because of the profligacy of its energy consumption. I was astonished at the results of the following exercise: if everybody on the planet consumed oil at the same rate as the average American, the industry would have to produce almost 450 million barrels a day, compared to about 81 million barrels a day in 2005, and the world's proved reserves would be exhausted in just over seven years.[21] By contrast, at the average consumption rate in Britain, global reserves would be spent in 17 years; Hungary, 34; China, 96; and India, 226. This disparity is the measure not only of America's greed and power, but also of its vulnerability. An entire economy built on this level of consumption cannot possibly survive the last oil shock, especially as several of the pillars on which it is based are already cracking.

The indigenous American car manufacturing industry for example, after fifty years of decline, is now already flirting with bankruptcy. Despite downsizing brutally during every recession since the early 1980s, today US carmakers are in deeper trouble than ever, as nimble Japanese competitors continue to steal market share from under their bloated SUV tyres. Whereas in 1955 the 'big three' US carmakers – General Motors, Ford and Chrysler – made 95 per cent of the cars sold in America, today their share is just 58 per cent and still falling hard. And while Honda, Toyota and Nissan achieved record-breaking sales in the US in 2005, General Motors and Ford together made losses of $7.2 billion, not to mention billions more in exceptional restructuring charges.[22] They now seem doomed to suffer the inevitable consequences of a strategy that backed their business model into a competitive cul-de-sac that the last oil shock is about to obliterate for ever.

The seeds of their downfall were sown, as you might have guessed, in a row about chickens. During a trade spat in the 1960s, Europe imposed import duties on frozen birds from America, and the US retaliated by slapping a 25 per cent tax on all imported light trucks, regardless of the country of origin.[23] When relations eventually thawed, the trade in poultry was un-frozen, but unaccountably the truck duties remained. So when Japanese manufacturers began to make inroads into the US car market in the 1970s the import tariff gave American carmakers a protected niche in which to hide from the new competition. While the Japanese took an ever increasing share of the US car market, the Americans could at least keep this profitable segment all to themselves.

This rank protectionism might not have mattered but for the impact of the first oil shock in 1973. As the oil price soared, Congress introduced a law that ordered each US carmaker to double the average mileage achieved by its vehicles to 27mpg by 1985. Light trucks were excluded from the new rules – known as the CAFE, or Corporate Average Fuel Economy standards – which made some limited sense for pick-up trucks and other work vehicles designed for hauling heavy loads. But incredibly the industry was also able to persuade Congress to classify some of the biggest and heaviest passenger vehicles, such as the Ford Bronco and Chevrolet Blazer, as light trucks, despite the fact they were designed to haul only people.

It was a loophole through which the industry would drive something rather less eco-friendly than a coach and horses. Over the decades US carmakers have spent billions of dollars – $9 billion in the 1990s alone – advertising SUVs as symbols of power, status and freedom, so that by 2004 they accounted for around a quarter of all American 'light' vehicle sales.[24] As a result, the US passenger fleet as a whole consumed far more fuel than it otherwise would have, and emitted far more carbon dioxide. None of this troubled the US car industry, because for a time at least the SUV was their one reliable source of profit.

'Profit' may not be exactly the right word, however, since it was effectively the result of massive state subsidies. SUV sales would never have exploded as they have without a lengthening list of absurd tax and regulatory distortions; the import tariff and exemption from the CAFE standards were just for starters. The SUV is also sustained by the negligible duty levied on US petrol, exemption from a 10 per cent tax on luxury cars, and a provision introduced by the Bush administration in 2003 which allows self-employed Americans to write off

$100,000 spent on an SUV against income tax – provided the beast weighs more than 6,000 pounds (almost three tonnes).[25] In the supposed home of free enterprise, this market is utterly rigged by the state to encourage the use of the heaviest and most destructive vehicles. It may just be coincidence, but during the 2004 election cycle the industry gave the Republicans almost $16 million dollars.[26]

But today even huge state subsidies may not be enough to keep the indigenous American car industry afloat much longer. General Motors and Ford in particular stagger under the weight of the healthcare and pension costs of all the workers they've already shed, while the non-unionized Japanese are now so efficient they can even compete in the SUV sector *despite* their 25 per cent penalty. Most importantly of all, hurricanes Katrina and Rita showed that the SUV market is indeed sensitive to the oil price. Sales of the 13mpg Ford Explorer slumped by 29 per cent in 2005, while those of the super-efficient Toyota Prius doubled to 108,000.[27] The hybrid electric car represents only a minuscule fraction of the total market so far – I didn't see many in Texas – but for the makers of the American gas-guzzler these figures should be as clear as a stop light.

There are plenty of other signals too. General Motors and Ford plan to cut another 60,000 jobs and twenty-six plants by the end of the decade; both companies' bonds are now rated as 'junk' on the debt markets, meaning they have the credit rating of a banana republic; and at one point General Motors – the world's biggest carmaker by production volume – was worth less than Harley-Davidson.[28] Like the lumbering dinosaurs their SUVs resemble, these companies have evolved into a state-subsidized Darwinian dead-end, and the last oil shock may be the asteroid that finally finishes them off. And when it does the repercussions will be pronounced: every 100 jobs in US car manufacturing support 460 others in industries such as steel, rubber, glass and electronics.[29] What is good for General Motors is no longer good for America, still less the world.

If anything, America's airlines are in even worse shape than its carmakers. The global airline industry has been in deep recession since about the turn of the century, losing *$44 billion* in the five years between 2001 and 2005, which according to British Airways' former chief executive Rod Eddington is more money than the industry has made in its entire history.[30] It makes you wonder why they bother. Like the carmakers, America's big established airlines are under attack from fierce new competitors – the low-cost airlines – and saddled with a legacy of pension promises and wage bills they can no longer support. Several, like Delta and Northwest, are trapped in chapter 11 bankruptcy proceedings, where it seems that every time they manage to force their staff to accept another massive pay cut, the oil price rises again to put the restructuring plan back in the red. The International Air Transport Association predicts the industry as a whole will return to profitability, but only on the basis of a far-fetched forecast of crude oil at just $52 a barrel.[31]

When the last oil shock hits the oil price will go in quite another direction. At, say, $150 a barrel America's legacy airlines will drop like flies, as will others around the world, to be followed at a short distance by their 'low-cost' rivals. As with the car industry, aviation redundancies will ricochet around the economy, through aerospace manufacturing, service companies, travel agencies, airport retailing, and banking: according to one industry estimate, 40 per cent of Citigroup's credit card revenues are generated by American Airlines' frequent flyer programme.[32] And remember, these are simply the *price* effects of the last oil shock. Once the oil starts to run physically short, there is nothing that can readily replace jet fuel except synthetic kerosene made by the Fischer-Tropsch process using coal or gas. These plants are expensive, slow to build, terrible for the climate, and as we saw in chapter 6 much of the gas is already spoken for.

If transport is under threat from the last oil shock, so is everything else that depends on it, including the entire American suburban economic landscape, and its acme, Wal-Mart. The world's largest retailer is often criticized for its treatment of staff, suppliers and the environment, but its shareholders will soon have rather more to worry about: the demolition of its business model. Unlike the carmakers and airlines, Wal-Mart is still making money – and lots of it. But its 'big-box' stores sited in the middle of nowhere, with their pile 'em high and sell 'em cheap approach, are completely dependent on a continuing supply of affordable fuel, and are therefore highly vulnerable to the last oil shock.

The only way for Americans to shop at Wal-Mart of course is to drive there, and its predominantly low-to-middle-income customers are precisely those whose spending power will be eaten into first by rising petrol prices. In fact this is already beginning to happen: in August 2005 Wal-Mart management blamed disappointing sales figures on the rising oil price, which also raised the company's fuel bill by $100 million in one three-month period.[33] If this was the result of $50 oil, imagine the impact of crude at $150 or $200.

At the other end of the business, Wal-Mart sources its goods overwhelmingly from China, at a cost of $18 billion in 2004, about one-tenth of America's trade deficit with the People's Republic. If it were a country, Wal-Mart would be China's eighth largest trading partner, ahead of Russia, Australia and Canada.[34] All these goods have to be shipped thousands of miles, making the company even more vulnerable to an oil price spike. Throw in a dollar collapse and Wal-Mart and its peers will find their costs soaring as their incomes plunge. Global trade will shrivel in the heat of the last oil shock.

As mall assistants join car workers and cabin crew in the unemployment lines, the American recession will gather force. And as people drive and consume less all sorts of businesses will go under. The housing bubble will burst as families who have 'bought' houses and cars with 100 per cent loans suffer redundancies and find their homes and wheels repossessed. 'Real estate must tank,' Robert Hirsch told me, before it actually began to do so in 2006, 'especially real estate that is further away from cities and jobs. You're talking about people literally being out on the streets.'

The severity of the downturn will hobble attempts to mitigate both the last oil shock and climate change. Redundant workers, even if they are lucky enough to leave with a pay-off, will naturally eke it out for survival rather than blow it on a fuel-efficient car, double glazing or any other kind of energy-saving investment. As usual the worst off will be least able to adapt. The same logic applies to companies and eventually, as tax revenues plummet, to the federal government. And as the prospect of economic growth becomes ever more distant, the traditional Keynesian strategy of borrowing your way out of a downturn may no longer be available; which sane investor would lend the money when there is so little prospect of the state ever being able to repay? In short, the last oil shock will mean the end of the US economic model, and if there is any justice George Bush's declaration that the 'American way of life is not negotiable' will be engraved on his tombstone as famous last words.

Any European urge to relish *schadenfreude* at the consequences of America's grotesque energy consumption should be resisted however. US consumer demand has been the flywheel of the world economy – so in one sense we all benefit and are all culpable – and when its axle snaps we will all suffer. And in any event, in the grand scheme of things European levels of oil consumption and dependency are different from America's only in degree, not in kind.

Having complained at the outset about the inadequacy of the term *peak oil,* I now have to confess that in some respects even *last oil shock* is not up to the job; the crisis will not stop at oil, and will quickly infect other energy markets such as natural gas. As we saw in chapter 6, there is no immediate shortage of gas underground – the resource is about 25 per cent depleted by most estimates of ultimate – but that's not the point. Oil and gas are closely intertwined through financial markets and geopolitics, and as a result when the last oil shock breaks the effect on gas is likely to be severe.

There will be upward pressure on natural gas prices in any event. One reason is likely to be a continuing lack of any meaningful spare production capacity. The industry will build significant additional gas capacity in the next few years, but will struggle to keep pace with strong demand growth and depletion. As we saw in chapter 6, the International Energy Agency clearly doubts that investment in the key Middle Eastern and North African countries will be sufficient to meet even the traditional sources of gas demand growth.

The advent of liquefied natural gas may also paradoxically help to keep gas prices higher. Rising supplies of LNG are not expected to ease the market significantly until 2015 or later, according the Deutsche Bank analyst Paul Sankey.[35] However the growing international trade in LNG will mean that the gas market becomes less regional and increasingly global, so that a growing shortage in one country will tend to be reflected in higher prices everywhere. According to the Ziff energy consultancy, for example, Canadian gas exports to the US will drop by 27 per cent over the next ten years.[36] If that leads to a sharp rise in American demand for imported LNG, then the price of cargoes bound for Europe will also rise. LNG tankers will steam to wherever the price is highest, whether that's Louisiana, Spain, or the Isle of Grain in Kent. From now on a gas shortage in America will raise prices in Britain and vice versa.

Yet another threat to the gas price comes from the fact that Russia, quite unexpectedly, turns out to have won the Cold War. It may have been forced to ditch its Soviet ideology, but of the three blocs Russia alone has both the nuclear weapons *and* the oil and gas. China and the West, by contrast, are now competing supplicants for Russian resources, giving enormous power to Moscow. President Putin demonstrated just how ruthlessly he is prepared to exploit this new-found strength when Gazprom severed supplies to Ukraine in the depths of winter to enforce a doubling of the price.

Since Europe receives 25 per cent of its gas from Russia, much of it through the Ukrainian pipeline, countries throughout the enlarged EU were the unintended casualties of the dispute. To Moscow's fury and embarrassment Ukraine kept drawing gas from the system which Gazprom intended for countries further downstream. A shortfall of up to 40 per cent was felt as far west as Germany and France, and Britain only narrowly escaped because it was still – just – a net exporter of gas. North Sea gas production is plummeting, however, so if it ever happens again, Britain will not escape a second time.

Although the impact on EU countries was unintentional, no sooner had supplies been restored than Russia began to threaten Europe explicitly. When talk of a Gazprom bid for UK gas supplier Centrica prompted speculation in the British press that any such bid would be blocked by the government, Gazprom chairman Alexei Miller bluntly explained the new rules of the game: 'Attempts to limit Gazprom's activities in the European market . . . will not produce good results,' he told a gathering of European ambassadors in Moscow. And to rising consternation: 'It should not be forgotten that we are actively seeking new markets such as North America and China. It's no coincidence that competition for energy resources is growing.'[37]

Miller was not just talking tough. Only the previous month Putin had signed a deal with China's President Hu Jintao to build two gas pipelines to supply the People's Republic, each with a capacity of up to 40 billion cubic metres of gas a year.[38] In other words, each pipe could deliver almost twice the annual consumption of Brazil. Worse still, the gas for the first pipeline – scheduled to start operating in 2011 – will come from Western Siberia, the region that supplies all the gas Russia currently sells to Europe. So having made an explicit threat to play its customers off against each other, which would inevitably drive up the price, Russia will soon have the means to carry it out.

And as if Russia did not already have enough power over the European gas market, in August 2006 Gazprom signed a broad agreement with Sonatrach, the state-owned Algerian gas company which is Europe's second largest supplier, to work together on LNG projects and buy up foreign oil and gas assets. The deal caused particular consternation in Italy which is overwhelmingly dependent on Russian and Algerian gas imports, and fears a pincer movement to force prices higher. Some in the industry even see this development as the beginning of a gas cartel on the lines of OPEC.

There are plenty of reasons for the gas price to rise of its own volition, but when the last oil shock breaks there will be another, even more significant: the tightening relationship between oil and gas prices on financial markets. Historically the two have always been connected, because many wholesale gas supply contracts stipulate that the price paid for the gas will be calculated by some formula related to the oil price. This is particularly true in Europe, and since Britain will import ever more gas from the Continent as North Sea production continues to fall, the price of oil will increasingly determine gas prices in the UK.

Because of this the British government is pushing Europe to reform gas pricing in order to separate it from the oil price. But even if the reform goes ahead, the two prices are unlikely to decouple. For one thing all the new LNG supply contracts from places such as Qatar are explicitly linked to the oil price, and for very good reason: it protects producing countries, and the banks that finance the construction of their multi-billion-dollar LNG projects, against any fall in the value of the dollar. Given the precariousness of the currency, this insurance of their future income is now more valuable than ever, and they are unlikely to give it up. Second, even in markets such as the United States where gas supply contracts are *not* explicitly linked to the oil price, traders on the financial markets behave increasingly as if they were. The US has $60 billion invested in hedge funds that make money playing commodities off against each other, and this means prices of oil and gas tend to move in tandem. And third, as oil supplies become scarcer, one obvious response will be to use natural gas as a transport fuel where possible, in the form of compressed natural gas (CNG) or synthetic diesel (GTL). In short, the price of gas is likely to become more tightly bound to oil, not less.

So when the last oil shock breaks, and the oil price spikes to unprecedented levels, gas is likely to go with it. And since gas is so prominent in power generation, so will the wholesale price of electricity. In this way the last oil shock will become an all-embracing energy crisis, long before supplies of other fuels begin to run short.

9

Memo to Mr Wicks

THERE WAS GOOD news and bad news. The good news was that at least one official working for the British government had a basic grasp of oil depletion and its importance. The bad news was that it wasn't the minister nominally in charge of UK energy policy. I discovered this on 12 July 2005, when both men were giving speeches in London – although at separate events – and each took a few questions from his audience afterwards. In each case I simply asked, 'When do you expect worldwide oil production to peak and go into terminal decline, and does it matter?' The answers I got were radically different, very illuminating, and worth quoting at some length. The good news – such as it is – I will save for later.

The dismal news came from the mouth of Malcolm Wicks who, upon his appointment as energy minister in May 2005, was described in the press by one starry-eyed anonymous source as a 'brain on stilts'.[1] They may like to reconsider after reading the following:

> Well, I always like to help a struggling author with a book. When? Well actually August 29th, I can't tell you what year though. Clearly fossil fuels are finite, aren't they – isn't that by definition the case – and there will come a time when there's no more coal, no more oil, and no more gas. I would have thought by definition that is the case. But I think when that August day approaches is a matter of great controversy, and it's a bit like if you look at our home patch, own back yard the North Sea, clearly at the moment if you look at the line on the graph paper we're exploiting that less and less successfully, yes? It is in decline, but when you talk to the oil companies and other experts exactly when production will end is unclear. I was out on the Elgin-Franklin platform a few weeks ago and that's going to be there for another twenty or twenty-five years. So the serious answer to your question is that none of us know when the oil is

going to run out, but it's not in the foreseeable future . . . But when it's going to run out, do you know, can you tell us? I mean I don't know.[2]

Leaving aside the gratuitous put-down and the generally fatuous tone, Mr Wicks's reply demonstrated apparently bullet-proof ignorance of the basics of oil depletion. Having been asked about when oil production will *peak* he chose to talk about when it will *run out*, which is usually the clearest sign that somebody just doesn't get it. The date when production finally comes to an end is almost completely irrelevant, since by that stage the amount of oil being produced will be negligible. We will already have had to learn to cope with a massive decline in output, and if we haven't, God help us. So Mr Wicks's use of the phrase suggests either that he truly doesn't understand, or that he was obfuscating, in which case you have to ask yourself why.

He certainly seems hostile to the idea that any forecasting effort could ever be worth while – 'August 29th' was clearly intended to ridicule. And yet despite maintaining that 'none of us know when the oil is going to run out' he is quite confident that it is 'not for the foreseeable future'. This 'serious answer' is both a contradiction in terms – how can you be sure if you can't forecast a date? – and at the same time 'a beautiful case of prediction by ambiguous statement', to borrow Hubbert's phrase. Based on *what* exactly, Mr Wicks?

It is interesting that he cited the UK North Sea while stubbornly refusing to draw its most obvious lesson: that it has peaked, like sixty other oil-producing countries, and that the world must at some point follow the same pattern. He might also have mentioned that by July 2005 when he made these remarks North Sea production had already slumped by 30 per cent in just six years.[3] At that rate of decline output in 2015 will be barely a quarter of its peak level.

'Can you tell us?' Very funny, Mr Wicks. I'm a journalist not a scientist, but even I can see that once you eliminate the obviously flawed or biased contributions to the debate, there is a surprisingly high level of agreement among the serious forecasters about when the turning point will arrive. You may no longer care about the answer to the question, Mr Wicks, because in November 2006 you were reshuffled, your old job was abolished, and responsibility for energy handed to your boss the Trade Secretary. One can only hope this is a positive sign. But since you asked, and since I am always happy to help a junior minister who struggles with his brief, here is your idiot's guide to when worldwide oil production is likely to peak.

First let's deal with some misleading official pronouncements. Unlike the British government which offers only bland, unsubstantiated assurances of 'not for the foreseeable future', the US administration has bolstered its public position by forecasting a specific date for the global peak, but one that seems comfortably far off in the future. The Energy Information Administration (EIA), the statistical arm of the US Department of Energy, predicts a global peak for conventional oil in 2037, but can only do so by making some apparently absurd assumptions, clearly illustrated in figure 20.

The EIA forecast is based on a simple model using the United States Geological Survey's estimate that the world's ultimately recoverable conventional crude oil will amount to about 3 trillion barrels. As we saw in chapter 3, the USGS ultimate requires the world to discover 22 billion barrels per year, whereas since 1996 we have been discovering less than half that much, an average of 9 billion barrels per year, as the USGS itself has now admitted.[4] As a result the USGS ultimate should be deflated by at least 400 billion barrels, to around 2,600 billion barrels. But even using the original USGS figure, the EIA modellers apparently had to turn somersaults to avoid the conclusion that a conventional peak was likely in the next decade.[5]

In one scenario the EIA assumes an oil production growth rate of 2 per cent a year until half the USGS conventional oil estimate has been consumed, producing a peak in 2016 followed by a gradual decline. The resulting production profile looks perfectly plausible, and would be typical of any large oil-producing province such as the United States. But the EIA discards this profile in favour of one that is skewed dramatically to the right, in which the peak is far higher, and is delayed until an implausible three-quarters of the oil has been consumed, to be followed by a precipitous decline. This profile is unlike anything known to the oil industry. Since the EIA modellers appear to have ignored the evidence of their eyes, and given what we know about the advice being sought privately by Dick Cheney, this forecast of 2037 appears to be for public relations purposes only.

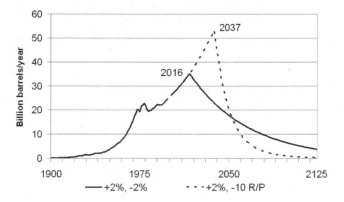

FIGURE 20. The EIA (US Department of Energy) forecast can only support a peak prediction of 2037 by adopting an improbable production profile

Like the American government, the International Energy Agency, the energy co-ordinating body for the OECD, maintains there will be no peak in global oil production before 2030, although it does now accept that non-OPEC output will peak before 2015.[6] The IEA's forecast is based on a complicated economic model that combines demand, fossil fuel supply, CO_2 emissions and investment. Unlike the EIA, it also covers total liquids production – not just conventional oil, but also oil sands, natural gas liquids, gas-to-liquids and so on. What the IEA model shares with that of the EIA, however, is a reliance on dubious data which fundamentally undermines its optimistic conclusion.

For OPEC the IEA uses IHS Energy's reserve figures, which although lower than the claims of the countries themselves, may still be overstated. If Colin Campbell is right, then OPEC's reserves are 250 billion barrels lower still. And for non-OPEC, the IEA model still relies on the unrevised USGS figure for ultimate. When I interviewed Nicola Pochettino, the energy analyst who created the supply side of the IEA model, he volunteered: 'If you want, this can be the weakness of the model. It's not our analysis, we took the data from outside and we assumed they are correct.' Having said this, however, Mr Pochettino did not appear to be aware that the USGS had already been forced to concede that discovery is running at less than half the rate required to fulfil the ultimate figure on which the IEA model relies.

Quite how much this affects the forecasts produced by the IEA model is not clear, but a potential shortfall of up to 650 billion barrels should have an enormous impact. The IEA model is further undermined by the open disbelief in its results expressed by some authorities in the Middle East. As we saw in chapter 7, both former and serving Saudi officials have disparaged the idea that the kingdom could ever produce the nearly 18 billion barrels per day that the IEA forecasts.

So the two most commonly cited official forecasts are based on a demonstrably extravagant resource estimate, and are therefore almost certainly too sanguine. It's true the big oil companies – with the honourable exception of Total – tend to concur with these assessments, but then they have shareholders to impress. Once you discount this group of forecasts, most of the rest fall in a spectrum so narrow that it could even be described as a consensus. Using a range of different approaches, these forecasters put the global peak no further away than 2020.

PFC Energy, the Washington-based firm consulted by Dick Cheney, produces the most optimistic credible forecast, but even this is pretty dispiriting. Their model is based not on traditional 'top down' Hubbertian analysis, where global discovery and production figures are extrapolated mathematically, but on a detailed 'bottom up' approach. First, production from every existing oilfield is modelled individually. Then the number, size and future production of oilfields yet to be discovered is projected by analysing historical trends. And finally everything is combined to produce a production forecast for each country and then the world. A range of oil price and technical scenarios are worked in, so that the results are presented with 90, 50 and 10 per cent probability outcomes.

PFC believes the most likely outcome is that non-OPEC production will peak in 2015 and that, depending on the rate of demand growth, global production will peak at 100 million barrels per day by 2020. PFC does not specify whether this peak will mark the start of a plateau or the onset of decline, but according to PFC's head of Upstream Economics Michael Rodgers, 'It's difficult to see a world where production capacity ever exceeds 100 mb/d.' Interestingly this was also the ceiling suggested to me in Vienna by the then Libyan Prime Minister Shokri Ghanem, who was formerly head of research at the OPEC secretariat.

Like most of the forecasts that follow, PFC discounts official OPEC reserves figures by a substantial margin, but it is important to stress that the company also deliberately builds in some elements of optimism. Rodgers explains: 'Our view is that if you always make optimistic assumptions and still end up with a problem, then you disarm some of the critics who claim you're crying wolf.' For example, the PFC model assumes that the exploration success rate of the last ten years will continue for the coming two decades – highly unlikely – and also adds another 20 per cent for reserves growth in existing fields. As a result the risks to the forecast are on the downside: 'The odds are pretty good that things will turn out worse.'

Despite BP's public hostility to the idea of an early peak, scientists at the company have produced their own models predicting just that. For some years Dr Richard Miller, a geologist and geochemist who works as an information manager in BP Exploration, circulated his forecasts within the company and to a small group of outsiders, but this stopped in 2005 after senior management asked him to withdraw a paper which had already been accepted by the peer-reviewed journal *Marine and Petroleum Geology*. Ian Vann, Group Vice-President for exploration and long-term renewal, disagreed with Miller's approach and told me he had not wanted BP's name attached to a forecast that 'seemed to be far too negative'. Miller concedes diplomatically that the paper could have caused embarrassment by contradicting the position of the company and its chief executive, but sticks by his analysis: 'At the end of the day the geological limitation will triumph over everything.'[7]

Like PFC, Miller took a field-by-field approach, to which he added his own estimate of the oil remaining to be discovered which, at 350 billion barrels, was in about the middle of the range. He also added a growth factor to represent small annual improvements in oilfield technology. Miller foresaw some spare capacity emerging in the oil market before 2010, possibly leading to a temporary price collapse, but non-OPEC went into decline by 2012, and the overall peak was forecast for 2019. From that point onwards he estimated that average daily production would fall by about 2 million barrels each year.

BGR, the German government's Federal Institute of Geosciences and Resources, uses the original and simplest 'top down' Hubbert method. BGR geologists assess a wide range of sources to come up with their own estimate of ultimate, although because they rely on official OPEC numbers for the big five Middle East producers their figure of 2,670 billion barrels for conventional oil is arguably too high. Then they calculate the number of years at the *current* rate of production to the halfway point, which by this method is 2017.[8] Peter Gerling, the head of BGR's Energy Resources Unit, stresses that the date is a first approximation, and prefers to predict the peak for 'the second half of the next decade'.

A report produced by the French government's General Directorate of Energy and Raw Materials, DIREM, predicts a peak in 2013 or 2023 based on two different sets of assumptions.[9] The 2013 forecast assumes that no more oil is discovered after 2005, which is clearly too pessimistic, but the 2023 forecast depends on our discovering 20 billion barrels annually, more than twice the rate achieved over the last few years. Both forecasts also rely on the assumption that output from existing production will decline at just 3 per cent a year, whereas many of the supermajors and oilfield service companies reckon the underlying decline rate is about 5 per cent.

Michael Smith is an oil industry consultant whose company Energyfiles runs a field-by-field model. He produces a spread of forecasts based on different assumptions about the growth of oil production until the peak – with dates ranging from 2008 to 2020. The strength of output growth in recent years had made the earlier end of the range look more likely, but Smith now expects the rising oil price to suppress demand, pushing the peak date back to 2016. From then he expects a 'corrugated plateau' for about five years, in which price spikes depress demand and production repeatedly below production capacity, until outright decline takes over and accelerates to a bobsleigh speed of 5 per cent a year.

Jean Laherrère, the former head of exploration technique at Total, agrees that the oil price and recession may delay the date of the peak by a few years. He expects a bumpy plateau to start any time now, and decline to start from about 2015. So too does Henry Groppe, the wise old contemporary of Hubbert. Colin Campbell, whose resource estimates are among the most conservative, uses a relatively simple country-by-country Hubbertian production model to produce a global peak in 2010, and, more alarming still, a peak for all hydrocarbons including gas by 2015. Laherrère has the gas peak in 2025.

Chris Skrebowski, the editor of the *Petroleum Review*, takes a different tack again. Because of the uncertainty over global reserves figures, he chooses to model oil companies' planned production against forecast demand. The lead time on big new oil projects is so long – typically over six years from discovery to first production – that it is possible to estimate fairly closely how much additional output can be expected for years in advance. Balancing that against the estimated decline rate of existing oil production, and the IEA's forecast demand growth, it is possible to assess how long the expected production increases from known projects can keep the market supplied. What Skrebowski is calculating is not a purely geological peak, but the ability of the industry to deliver, which could fail sooner. Skrebowski judges that by this measure, 'The wheels come off in about 2011.'[10]

Hubbert's later forecasting techniques dispensed with the need for a prior guess of the amount of oil that would ever be produced. In this tradition, Professor Kenneth Deffeyes has developed a new twist on Hubbert's logistic curve approach to derive an ultimate of 2 trillion barrels, and forecast a total liquids peak at the end of 2005.[11] At the time of writing, it already looked as if this prediction was premature, but it's worth remembering that Hubbert's first use of the logistic curve produced a forecast of the American peak that was three years too early. Deffeyes may not be wrong by much.

Which of these forecasts is correct is impossible to say, but given the narrowness of the range, the question becomes almost academic. It is no surprise that there is a variety of dates for the global peak that do not precisely agree; the differences in analytical approach and the unreliability of the data guarantee it. But the areas of uncertainty are evaporating with every year that passes, meaning that the current crop of forecasts has much less chance of being wrong than some previous attempts to predict the global peak (chapter 3). Now the world is so well explored for oil there is less room for error.

The biggest remaining uncertainty is OPEC, and it is a big one. The OPEC peak is the outer bound for the global peak, since it is almost universally accepted that non-OPEC will already be in decline by the middle of the next decade at the latest. The global peak could however come sooner, if non-OPEC output falls faster than OPEC production can grow. OPEC may have bigger reserves than these forecasts assume, and may be able to grow its production more quickly, but given the extreme doubts surrounding its unsubstantiated official numbers, to plan our energy policy on that basis would be lunacy.

So, Mr Wicks, in answer to your cheap crack, 'When? Well actually August 29th, I can't tell you what year though', neither can I. But it doesn't actually matter, provided you recognize that the peak will happen, almost certainly within the next fifteen years, and probably sooner rather than later. And provided you recognize also that even if by some miracle oil production does not peak in that timeframe, we would be far better off assuming that it will. Your apparent ignorance is therefore terrifying; as Robert Hirsch's report for the US Department of Energy shows, we have little enough time to prepare whichever date eventually turns out to be correct.

But please don't take my word for it, Mr Wicks. On the same day I asked you, 'When do you expect worldwide oil production to peak and go into terminal decline, and does it matter?' I asked the same question of the government's chief scientific adviser, Sir David King. The good news, such as it is, was that his reply differed radically from yours:

> Well, this is a very good question, and as you know there are many books currently being published on the subject, and I have waded my way through many of them. This issue is in one sense relatively simple. It's just a question of when we have a peak. There is a finite resource of oil within the earth's surface and there must be some point at which we have depleted half of it, and after that point it's going to be rather difficult to maintain supplies in keeping with demand . . . I would have to say that within the next ten years we're going to see a situation where productive capacity no longer meets normal demand, in which case we're going to see a real demand for alternatives . . . I don't think it's going to appear as a Gaussian [a particular type of bell curve] peak, I think we're more likely to see a plateau . . . *But ten years or less* [My emphasis].[12]

You should have talked to your chief scientific adviser a bit more, Mr Wicks, perhaps borrowed some of his books. Or maybe had a chat with Sir David Manning, Britain's ambassador to Washington, who also seems to get it, on the evidence of this speech at Stanford University in 2006:

The supplies of oil on which we depend are finite. Global oil production is apparently nearing its peak. Although there is intense debate
about exactly when this will happen . . . current estimates seem to be
converging on some point between 2010 and 2020 . . .

The International Energy Agency predicts that, if we do nothing,
global oil demand will reach 121 million barrels per day by 2030, up
from 85 million barrels today. That will require increasing production
by 37 million barrels per day over the next 25 years, of which 25
million barrels per day has yet to be discovered. That is, we'll have to
find four petroleum systems that are each the size of the North Sea.

Is this realistic? Production from existing fields is dropping at about
5% per year. Only one barrel of oil is now being discovered for every
four consumed. Globally, the discovery rate of untapped oil peaked in
the late 1960s. Over the past decade, oil production has been falling in
33 of the world's 48 largest oil producing countries, including six of
the 11 members of OPEC. How then will we meet the soaring
demand that the growing global economy will require?[13]

Or had you ever felt the urge, Mr Wicks, you could have learned a lot
from this publicly available report from the US Army Corps of Engineers,
which again is in stark contrast to your stunning complacency:

The doubling of oil prices from 2003–2005 is not an anomaly, but a
picture of the future. Oil production is approaching its peak; low
growth in availability can be expected for the next 5 to 10 years. As
worldwide petroleum production peaks, geopolitics and market economics will cause even more significant price increases and security
risks. One can only speculate at the outcome from this scenario as
world petroleum production declines. The disruption of world oil
markets may also affect world natural gas markets since most of the
natural gas reserves are collocated with the oil reserves.[14]

In short, Mr Wicks, your apparent ignorance is inexcusable. It is also
something of a riddle. If, as I argue in chapter 1, an awareness of the
approaching peak was fundamental to the Anglo-American decision
to invade Iraq, how is it that you as energy minister appeared to know
so little of the energy facts of life? Perhaps the charitable answer is that
knowledge of the global peak within the British government is
encouraged on a need-to-know basis, and you didn't need to. After
all, the extreme centralizing tendencies of the Blair government are

well known, and if the Labour Party treasurer wasn't told about the cash that sparked the 'loans-for-honours' scandal, why should you have been expected to know about one of the most fundamental aspects of your brief?

To be fair, your brief was tightly drawn: to oversee the six-month 'Energy Review', that transparent softening-up exercise for the introduction of a new fleet of nuclear power stations which masqueraded as a consultation exercise, and delivered its largely foregone conclusions in the summer of 2006. During the preceding winter of gas shortages, you lost no opportunity to insinuate that Britain might be better off with 'cleaner home-grown sources of energy', although as far as I am aware we don't have any uranium mines in Britain.[15] The policy may be sensible, but the suggestion that a decision had not already been taken in Downing Street is absurd; Tony Blair couldn't even wait for the final report to announce the new policy in a speech to the CBI. If you weren't told when you took the job that the decision you would be promoting was driven both by climate change *and* the impending global peak, well, it wouldn't make any difference to your performance as PR man.

On the other hand both Sir David King and Sir David Manning are known to be close to Tony Blair, and the fact that they so clearly *get it* tends to reinforce the idea that the global peak has driven Britain's foreign policy, and what passes for its energy policy – even if the energy minister hadn't the foggiest idea of its significance. The two Sir Davids are the common threads that link British policies on Iraq, Iran and UK nuclear power: King believes nuclear renewal is a 'scientific necessity', Manning was Blair's foreign policy adviser during the crucial years between 9/11 and the invasion of Iraq, and both expect oil production to peak within the next decade or so.[16]

There may however be another explanation, one that does not rely solely on your ignorance, Mr Wicks, to explain the schizoid appearance of UK policy (**schizoid** /skitsoyd/ *adj*. Tending to resemble schizophrenia or a schizophrenic, *but usually without delusions*). It has to do with the enormity of the consequences of the last oil shock, and the fundamental difficulty that politicians of all stripes will have in confronting them. In other words, they may not be ignorant, but well informed and terrified.

The problem was sharply exposed in the spring of 2006 when Gordon Brown and David Cameron started competing farcically for the 'green' vote. 'Dave' invited ridicule by cycling to work – with his chauffeur-driven car following on behind; Gordon by implying when interviewed that we could save the planet if only we all remembered to turn the telly off at night. Worse still, both insisted in almost identical terms that nothing need affect our living standards. Cameron coined the disingenuous catchphrase 'green growth' and declared: 'We have to liberate ourselves from the myth that we have to choose between protecting the environment and promoting prosperity.' Brown advocated 'the new synthesis', claiming that 'far from being at odds with each other, our economic objectives and our environmental objectives now increasingly reinforce each other'.[17] Both men are highly intelligent and neither can possibly believe the fantasy that greenhouse gas emissions can be cut by 60 per cent by the middle of the century without a massive impact on consumption and economic growth. But which pretender could expect to inherit Blair's crown if they were to spell out the scale of the sacrifices that will be required? In an excoriating piece, the *Observer* columnist Andrew Rawnsley summed up: 'The politician who means it is the one who tells you where and how much it is going to hurt. David Cameron and Gordon Brown are both still suggesting that we can have our planet and eat it.'[18]

This is not simply the failing of two individual politicians, but of democracy itself – a system in which parties compete for votes among an electorate largely ignorant of the scale of the problem, and kept in ignorance by politicians' desire to win power. (I'm not suggesting any alternative system, but this is a necessary effect of the one we have.) And it's a contradiction that is glaringly enshrined in Blair's own position: he came up with Britain's ambitious emissions reduction targets, and yet pursues a deliberate policy of unbridled expansion of air travel – jeering at fellow MPs in committee if they thought there were any votes in making flights more expensive.[19] He clearly understands, as do Brown and Cameron, that no politician can expect to get elected on a programme of belt-tightening, so the necessary domestic action is deferred, and any politician serious about power must continue to lie. And for those who hold power, launching resource wars abroad in a vain attempt to defer the global peak may be preferable to policies that hit lifestyles and polling-day prospects at home.

Perhaps then, Mr Wicks, your ignorance stems from the conflicted position of your Department of Trade and Industry, whose Statement of Purpose is 'creating the conditions for business success'. How would that be achieved if it was accepted that, in addition to the challenges of climate change, global oil supplies could soon start to fall at up to 5 per cent a year? Wouldn't this confront the bureaucrats at the DTI with some seriously existential questions? And perhaps this in turn explains the department's long-established aversion to thinking about when global oil production will peak.

Unlike its French and German counterparts, the DTI has never produced its own forecast of the likely date, despite the mounting evidence, and despite the fact that academics at Reading University have been trying to persuade them to take the issue seriously for a decade. Senior Research Fellow Roger Bentley organized an informal group of scientists at the university to assess the various forecasting approaches in 1995, and soon realized the issue was so serious it demanded government attention. A number of meetings were held with DTI officials over the years, but on each occasion their approach was rebuffed. 'I don't mind if they disagree with us,' says Bentley, 'but what is truly wicked is their refusal even to put a couple of scientists on to it for a couple of months just to see whether the thing is worth looking into. That's just criminal.'[20]

Despite your glib dismissals, Mr Wicks, your officials were subsequently forced to respond to further questions about the oil peak, although their arguments remain signally unconvincing. In early 2006 James Howard of the peak oil pressure group PowerSwitch wrote to you demanding to know what the UK government's position was, and what you proposed to do about it. This is part of the reply your officials drafted for you:

> The Government is aware of the arguments surrounding this issue that global oil (and gas) production will one day peak, which cannot be disputed. However, we believe that such a peak is not imminent and will not be reached until some time after 2030, provided the necessary investments in expanding and replacing production capacity are made.
>
> The International Energy Agency is currently developing detailed medium-term (to 2010) projections for global oil production based on

field-by-field analysis of active development projects and existing decline rates . . . Rather than showing any imminent decline in production, these findings showed global oil production capacity rising steadily to 2010 (by around 2 million barrels per day each year).

Further into the future, new oil discoveries will be needed to renew reserves. While discoveries of new oilfields have fallen sharply since the 1960s, this fall has been most dramatic in the Middle East and the former Soviet Union and is largely the result of reduced exploration activity in those regions with the largest reserves, and also of a fall in the average size of fields discovered. Exploration drilling in the Middle East has been minimal for many years because existing proven reserves are very large. Rather, drilling has been concentrated in North America and Europe, which are both mature regions. However, we are already seeing signs of a rebound in exploration and appraisal drilling in the Middle East, which can be expected to accelerate in the coming decades. Not only will this help to renew reserves, but it will also increase the average size of fields discovered. In addition to the Middle East and former Soviet Union, North and West Africa, deep-water Gulf of Mexico, and Latin America are also regions where new oil discoveries can be expected.[21]

The first paragraph I will admit contains a minuscule advance – finally, an official acknowledgement that the peak is inevitable. But the idea that it won't happen until 2030 is evidently based on the IEA's long-range forecasts, which as we have seen are based in turn on USGS numbers that have already been shown to be a substantial overestimate.

Paragraph two is hardly reassuring, asserting only that oil production will continue to rise for the next five years, a judgement with which most peak forecasters agree. The IEA's Medium Term Oil Report to which this paragraph refers has since been published, a close reading of which gives a very different impression to the one implied here. While it does forecast rising capacity to 2010, it concludes that by 2011 'capacity growth is already seen slowing and demand will have strengthened', and on that cliffhanger the IEA medium-term forecast ends. Remember, 2011 is the year in which Chris Skrebowski's analysis suggests that supply will first fail to meet forecast demand. It must have been a great relief to know it would not happen in your term of office, Mr Wicks, but what then?

It is paragraph three, however, that contains the real howlers – all the articles of economic faith and self-delusion strung together to produce a reassuring fairy tale. I forwarded your letter to Michael Rodgers, head of Upstream Economics for PFC Energy and a former oil explorer himself, and here's what he wrote back:

> This kind of argument makes me crazy. The fact is that oilfield sizes have fallen everywhere and there has been a lot of exploration outside of Europe and North America after the 1960s. People who use the argument that field sizes have fallen because there has been less explo-ration assume that a company deliberately looks for smaller fields when they have less exploration dollars to spend. The argument goes exactly the other way . . . as an explorationist when you know you have fewer wells to drill you are more careful than ever to make sure that what you drill has the most reserve potential and lowest risk. The last two years of high oil prices correspond to the worst exploration results on record since WWII. Think about that for a second . . . the last time explo-ration results were worse than the combined years of 2004 and 2005 was during WWII when you could not do any exploration . . . Everybody sees the Middle East as the solution but in actual fact there has been exploration there over the last 20 years and field sizes there have fallen just like everywhere else.

No wonder you were so badly briefed, Mr Wicks. Not only were you evidently out of the Number 10 loop, not only were you the *seventh* here-today-gone-tomorrow energy minister in less than a decade of New Labour rule, but DTI officials remain obstinately immune to the reality of peak oil, and resist all encouragement to investigate the idea. Perhaps they are card-carrying economists, perhaps they are in deep denial, or perhaps they understand the gravity of the impending crisis perfectly well, but are paralysed by the scale of the challenge.

10

Passnotes for Policymakers

THE SCALE OF the challenge posed by the last oil shock is certainly enormous, and the potential for planetary catastrophe high, particularly as it comes in a one-two combination with climate change. Nevertheless, I remain surprisingly optimistic about our chances of surviving it in reasonable order. There are no silver bullets, and I am certain that our consumption of energy will fall massively in the coming decades, but this need not be the end of our civilization. Many of the technologies and ideas needed to combat the crisis already exist, and what is required is political and popular will to engage them. With any luck the early tremors of the last oil shock will concentrate minds, blow away the denial and disingenuity, and prompt a serious and co-operative response at both the national and international level. In any event, there is a great deal that individuals can do in their own lives to protect themselves and their families from what lies ahead. And although life will certainly be very different and more constrained after the last oil shock, in some ways it may even be better.

At some point in the next decade or so, the world's oil supply will start to fall. Non-conventional oil – though huge in resource terms – will not expand fast enough to compensate for the decline in conventional. Nor will the much-touted 'alternatives' fill the deficit: hydrogen is a non-starter, certainly in terms of private transport; the potential of biofuels – in the developing world at least – is far greater, but still inadequate, and fraught with ecological risks. In short, we will soon face a growing liquid fuels shortage, and for the first time the oil supply will become a zero-sum game: as the pie shrinks, if one person or country consumes more, somebody else will have even less. We will have to co-operate, or fight.

At the same time, we have to tackle climate change; peak oil does not solve global warming. Even if oil production halves by 2030, the predicted increase in gas and coal consumption means that overall emissions would still rise. So while nature is likely to self-correct in terms of oil emissions, humanity has to find a way to achieve massive cuts in coal and gas emissions voluntarily, *and* adjust to a growing deficit in transport fuel.

Unless we are extraordinarily lucky, all this will have to be achieved during the throes of the last oil shock. As I write it looks as if we are already ratcheting into it. Spikes in the price of oil, whether caused by 'natural' disasters or international conflict, will yank the price of gas and electricity higher. These combined spikes may cause deep recessions or even a depression, which would undermine our ability to finance the necessary changes. Prices will be volatile, so the traditional economic signals will be confusing. Policies based on taxation or price alone will be overkill when oil prices soar, and ineffective when they slump. We will need a wholly new approach to economics.

This fundamental change will also have to be achieved from within a political, financial and economic system that is predicated on growth, and in which huge vested interests will resist the necessary pre-emptive action, which demands a recognition that *business as usual* means collective suicide. Despite the recent history, some political leaders will no doubt continue to prefer foreign resource adventures over confronting electorates at home with the vote-threatening truth.

The last oil shock undoubtedly presents a historic challenge to humanity. But although the obstacles are formidable, the broad route we need to take is quite clear. It goes without saying that emissions from gas and coal use must be cut hard, and the only debate is about how to achieve this. It is equally clear that we must also begin to reduce oil consumption right away. This might seem an odd suggestion at first: why not party on, guilt-free, for another few years, since nature is about to call time in any case? But there are very good reasons for pre-emptive action.

First, reducing consumption is a far more effective way of deferring the peak than hoping in vain for some massive new oil discovery. Thierry Desmarest, chief executive of Total, estimates that if oil demand growth were simply halved, the global oil peak which he expects in about 2020 would be delayed by a decade.[1] Desmarest was talking at a conference in Amsterdam in 2006, and went on to make what was for a senior oilman an astonishing plea: 'We say to governments, it's urgent to take action plans to reduce oil demand growth.' He was absolutely right: reducing consumption now would buy us time, and make the post-peak decline more gradual.

Second, when global production decline sets in, the change will be wrenchingly worse if we have not prepared for it. To the extent that we can achieve cutbacks pre-emptively, we may retain a measure of control and perhaps avoid a paralysing crisis. Many countries and individuals will almost certainly refuse to do so, on the basis that it would harm their economy or lifestyle, but when the peak comes, those who are already accustomed to cutting their oil consumption will cope much better. Since we have been put on notice that we are about to run a marathon, the winners are likely to be those who put in the training, even if it means bearing some pain beforehand. Those who do not prepare increase their risk of injury, and possibly heart attack.

At the moment of course all the trends are going in the wrong direction. Globally we are consuming more energy of all kinds and emitting more greenhouse gases with every year that passes. So the question is which technologies and policies could best achieve both substantial and continuing cuts in greenhouse gas emissions, and at the same time ameliorate the effects of the last oil shock, either by softening its near-term, market-driven impacts, or by providing any measure of additional energy to replace liquid transport fuels. Any policy should ideally satisfy both climate change and peak oil criteria, or, at the very least, no measure should be adopted to tackle one problem that would worsen the other. There is however a serious risk that this will in fact happen: falling conventional oil supplies will spur ever greater investment in coal-to-liquids production, which is filthy in terms of CO_2 emissions; and some of the policies proposed to tackle climate change may unwittingly increase our vulnerability to the last oil shock.

As soon as it is accepted that the oil peak is imminent, and that pre-emptive cuts in consumption are essential, a number of obvious practical measures immediately suggest themselves:

1. Launch a massive public education campaign; the radical changes that are needed will only be achieved if people truly understand the gravity of the crisis. And by educating the public, politicians would begin to get themselves off the hook.

2. Scrap all airport and road network expansion forthwith; there will be plenty of spare capacity soon enough.

3. Legislate or strengthen vehicle fuel efficiency standards, and/or institute tax measures to drive gas-guzzlers out of existence, or at least production. Ironically the US CAFE standards might provide a model – though with much stiffer targets. Alternatively, Amory Lovins of the Rocky Mountain Institute has proposed a system of 'feebates' in which those who insist on buying fuel-profligate vehicles would have to pay an extra fee – perhaps thousands of dollars, pounds or euros – to cut the price of the most fuel-efficient vehicles. The selfish would subsidize the sensible. Whatever the policy instruments, it must be made abundantly clear that there will be no market for vehicles such as the 18mpg Range Rover Sport within a very few years.

4. Secure international or at least European agreement to remove the current exemption of jet fuel from duty and VAT, and to impose VAT on aircraft and ticket sales. Consumers will howl, and airlines may go bust sooner than they otherwise would, but this loophole must be closed.

5. Divert any money saved or raised by points 2–4 into assessing and filling the gaps in the public transport network.

These piecemeal measures would be just the start of any programme to deal with the oil peak – necessary but nothing like sufficient. But the chances of getting even this much legislated are vanishingly slight as things stand, both because of the politicians' dilemma described in the last chapter, and because the mainstream energy debate continues studiously to ignore the issue. These two factors are of course closely related.

When the government published its hastily assembled Energy Review in July 2006 the contents had been well flagged and there were no major surprises. The Review had been framed solely in terms of climate change and 'energy security', and was overwhelmingly concerned with Britain's rapidly rising dependency on imported oil and gas, and the fact that a quarter (18 gigawatts (GW)) of our existing electricity generating capacity will be lost by 2023 with the closure of obsolete coal and nuclear power stations.

The tone was very different from the previous White Paper of 2003 which had complacently maintained that the impending shortfall could be made good with a mixture of renewables, efficiency gains and new gas-fired power stations. Now, all of a sudden, the government had apparently woken up to Britain's looming import dependency and self-inflicted electricity deficit. Environmentalists who approved of the earlier policy complained that they couldn't see what had changed in the interim. Part of the answer, I would argue, was a growing realization of the imminence of the global peak, at least in Number 10 if not in the DTI. In any event, the fact that this second review of energy policy was required within three years, and the speed with which it was conducted, spoke of a government starting to panic.

It was clear from the outset that the whole exercise had been designed principally to soften up public opinion for the government's U-turn on nuclear power, a decision Tony Blair later admitted he had already taken before launching the 'consultation' process.[2] He couldn't even wait for the Review to be published before announcing the thrust of the new policy in a speech to the CBI in May 2006:

> The facts are stark. By 2025, if current policy is unchanged, there will be a dramatic gap on our targets to reduce CO_2 emissions; we will become heavily dependent on gas; and at the same time move from being 80/90% self-reliant in gas to 80/90% dependent on foreign imports, mostly from the Middle East and Africa and Russia.
> These facts put the replacement of nuclear power stations, a big push on renewables and a step-change on energy efficiency, engaging both business and consumers, back on the agenda with a vengeance.[3]

Neither Blair's speech, nor the Energy Review itself, nor the subsequent Stern Review on the economics of climate change made any mention of the oil peak, and the omission is suspicious. In the Energy Review there is just one oblique reference to depletion buried at the back of the 200-page document: 'Over the long term we must continue to make progress in reducing carbon emissions on a path consistent with our 2050 goal [to cut carbon emissions by 60 per cent]. We will also need to adjust to the global depletion of fossil fuels. The actions we need to take to address these *long term* challenges are closely linked' (my emphasis). The implication here is that there will be no global peak for several decades, reflecting the government's established public position of 'not before 2030'. But since we know that the Energy Review was driven by Tony Blair, that he has been taking tutorials from the Bush administration through the US-UK Energy Dialogue, and that two of his closest advisers believe the peak will arrive in 'ten years or less', it follows that the official position is probably not the one that informed the new policy. Britain's official position – like that of the US – is therefore now best seen as propaganda.

It is this that makes the Energy Review such a conflicted document. Evidently it was at least partly inspired by the 'ten years or less' position of Blair's advisers, but because of the government's public stance the Review could not address the peak explicitly. Even by its own lights the Review's stance on transport is particularly weak, and amounts to praying for salvation through deus ex machina: 'Securing a change in people's transport behaviour and their choice of transport is also not straightforward. In the longer term we expect the emergence of new technologies including hybrids, advanced biofuels and hydrogen to play a major part in reducing transport emissions post 2020.' In other words, the government's policy on this vital issue is to hope 'something will turn up' – although not any time soon – citing technologies that we already know are severely resource-constrained. But as the quote in the previous paragraph shows, the Review also notes that the actions needed to tackle climate change and oil and gas depletion are 'closely linked', and it is in this light that its main policy proposals should be interpreted.

At first glance the new strategy – nuclear, renewables and efficiency – seems to have little to do with the last oil shock; after all, you cannot put nuclear or wind power in your tank, at least not directly. But as we have seen, energy markets are intimately linked, and as a result a British policy ostensibly designed to tackle the power generation deficit could also mitigate the effect of oil depletion on transport – but only if pushed much further than is apparently intended. And rather to my surprise, the Review contains a number of other ideas that could also make a profound difference to how well we cope with the last oil shock, but again, only if pursued further, faster and more rigorously than suggested here. It is far from certain that the government or any other party has the political will to achieve it, since doing so would involve taking voters' toys away, but the glimmer of a policy is there.

As a general proposition, for example, reducing our dependency on natural gas as Tony Blair suggests would be an excellent idea. Gas already provides 40 per cent of our electricity, against 33 per cent from coal, 19 per cent from nuclear and a paltry 3.6 per cent from renewables. As ageing nuclear and coal power stations are decommissioned, the default position is that we build yet more gas-fired power stations and become yet more dependent on gas – possibly 60 per cent and beyond. And this is only part of the picture: although we burn a lot of gas for electricity generation, we use *more than twice as much again* for industrial and domestic purposes.[4]

In addition to the general points that the Review identified, there is also a litany of specific reasons to want to reduce our gas dependency as quickly as possible: Russia may cut off supplies again for political ends; the IEA believes that Russian gas investment may prove inadequate even to maintain supplies for its existing customers; Russia has signalled clearly that it intends to play Europe off against China, and eventually the US; Qatar's fabled North Field may turn out to be much smaller than we thought; when the last oil shock breaks, potentially devastating spikes in the price of oil will be matched by gas and therefore electricity; and most important of all, gas production will almost certainly peak within the lifetime of the next generation of power stations.

Jean Laherrère, the former head of exploration technique at Total, predicts a global gas peak in 2025, and, perhaps more importantly, regional peaks far sooner. North American gas production seems to have peaked in 2003, and Laherrère judges Europe will go over the top in 2010, with Russia following by 2020. As we have seen LNG is unlikely to relieve these regional peaks in the short term. What's more, post peak production decline rates are far steeper for gas fields than for oil, so the gas peaks will be that much more brutal. We therefore need to have weaned ourselves off natural gas entirely within two to three decades at the outside.

Reducing our gas dependency in electricity generation should be a policy goal in its own right, but such a strategy *might* also help to relieve the physical shortage of transport fuels after the oil peak, at least for a time. Most road vehicles could fairly easily be converted to run on compressed natural gas, and the fuel is particularly suitable for buses. So any policy that replaced gas-fired power stations with other forms of generation would in theory release fuel for use in transport. Alternatively, if generating capacity could be expanded faster than traditional sources of demand growth, any 'excess' could be used to run trams, trolley-buses and electric cars. Either way, to make any dent in the oil peak, Britain would have to expand non-CO_2 emitting generating capacity by far more than required to fill the self-inflicted energy gap.

However it seems that Tony Blair's three-pronged approach as expressed will struggle to achieve even its more limited goals. To start with, the new nuclear power stations needed to replace those that are due to close may not be built in time. An enquiry by the House of Commons Environmental Audit Committee found that given the years needed for design, safety vetting, planning approval and construction, the first new plant was unlikely to be completed until 2019 at the earliest.[5] And if a total of ten reactors were built in quick succession, to replace all the lost nuclear capacity, the last would come on stream in about 2030. The Energy Review promises 'radical change' to speed up the planning process, but it remains to be seen if this will prove decisive, particularly since the government insists the private sector will have to shoulder all the risk. There are also serious doubts about whether the industry has the engineering capacity to build all the reactors required by the global renaissance in time.[6] The Review provides no target dates, and even with the proposed reforms it is probably too late for nuclear capacity to replace itself seamlessly.

Opponents of nuclear power often enlist this fact to argue that it is not worth pursuing this option at all, even those who like former Environment Secretary Michael Meacher are convinced the oil peak is imminent and apocalyptic. This is a curious position, which acknowledges that we face a huge energy deficit, but proposes to make it bigger anyway. On the contrary, it would be better not to abandon nuclear but to find safe ways to speed up its self-replacement – just as the government suggests – and even to increase the target generating capacity. A gas peak in 2025 would be less traumatic if we were well on track to replacing the lost nuclear capacity, with plans to expand to say 40 per cent, than if we had none at all. France, with almost 80 per cent nuclear electricity, is already far better placed in this respect. It is not what you would call sustainable, but we have to deal with the legacy of nuclear waste in any case, and one more generation of reactors would certainly help us to ride out the combined oil and gas peaks.

About a third of British coal-fired capacity (8GW) must close even sooner than the nuclear power stations because of an EU environmental directive. But in theory we could continue to burn coal, guilt-and-emissions-free, by installing *carbon capture and storage* (CCS) technology. The idea of CCS is to trap CO_2 emitted from power stations, and store it safely away from the atmosphere by pumping it into depleted oil and gas reservoirs, where much of it came from in the first place. This is perfectly feasible using current technology, and of course many of the necessary pipelines are already in place. Since the UK's twenty worst polluters generate 20 per cent of the country's *total* CO_2 emissions, the equipment need only be fitted to relatively few power stations to make a big difference. According to rough calculations by Dr Sam Holloway of the British Geological Survey, if this were done, the oil and gas fields in the British North Sea have the potential to store some fifty years of those power stations' emissions. And if saline aquifers under the southern North Sea also proved leak-proof – which would need further geological investigation – they could have the potential to hold another 100 years' worth.[7] In other words we could carry on burning cheap coal at the present rate until the current proved reserves ran out, while emitting a fraction of the CO_2.[8]

The government emphasized the enormous potential of CCS in its 2003 White Paper, but having promised much has since delivered little. At the time of writing, there was not even a demonstration project in operation. The House of Commons Environmental Audit Committee lambasted the government's lack of progress as 'scandalous', and declared that any failure to seize this opportunity would be 'grossly negligent'.[9] The Energy Review once more extolled the benefits of CCS, but yet again seems longer on talk than action. The document made clear that progress on CCS was being held up by negotiations to secure the necessary international legal framework and recognition under the EU Emissions Trading Scheme, and research into its likely cost – which would certainly be high. The Review set no date for the introduction of CCS, and on the evidence of this document there can be no confidence that it will be available in time to offset the impact of the 2015 EU deadline.

The Energy Review reiterates the government's commitment to achieving 10 per cent electricity generation from renewables by 2010, and its 'aspiration' to double that by 2020, but here too its record is miserable. Other countries have set themselves much stiffer targets and are pursuing them far more dynamically. The Environmental Audit Committee report noted that of the fifteen pre-enlargement members of the European Union, only Greece had a lower proportion of renewable electricity generation. In 2004 renewables generated less than 4 per cent of Britain's power, and four-fifths of that was from landfill gas, which is not strictly 'renewable'; eliminating waste in the first place saves more energy than is gained by exploiting it. In fact, less than 0.5 per cent of Britain's electricity came from wind, with just over 1GW capacity installed. By contrast, Germany had seventeen times as much wind-generating capacity as Britain, and nearly *100 times* our solar capacity. In the same year Spain installed eight times as much new wind capacity as Britain, and has been so successful that it has doubled its renewable generation target to 20GW (15 per cent) by 2010. Denmark already generates 20 per cent of its electricity from wind alone, while the German state of Schleswig Holstein has plans to produce half its power from wind by the end of the decade.

The reason these shameful international comparisons exist is New Labour's slavish adherence to a market-based approach, whereas the countries that have been most successful at encouraging renewables are those that are unencumbered by this ideological fixation. Britain's policy is known as the Renewables Obligation, which obliges electricity suppliers to obtain a certain percentage of their power from renewable sources, through a relatively complicated system of tradable certificates. Germany, Denmark and Spain have all opted for a 'feed-in tariff', which amounts to a straight subsidy. The results speak for themselves.

The Energy Review focused on the potential for offshore wind, and rightly so, since its wind farms are far less likely to provoke public opposition and the resource potential in Britain is immense, just as it is for coastal Europe and the United States. A study conducted by the Central Electricity Generating Board in 1982 concluded that Britain could generate 360 terrawatt hours (TWh) of electricity per year from offshore wind, almost equal to our current consumption of 378TWh/year.[10] But one of the authors, David Milborrow, explains that while the study was rigorous, its assumptions about issues such as the spacing of wind turbines were very conservative compared to current industry practice. Today he judges the potential resource is double the original estimate, about 720TWh/year. Another report for the European Commission concluded the UK resource was larger still at 986TWh/year.[11] Either way, if British waters bristled with turbines, offshore wind alone could in theory cover all our electricity and a very sizeable chunk of the rest of our energy needs.

Talk of energy shortage amid such plenty begins to sound absurd, but the problem as usual is getting from here to there. So far Britain has built only four offshore wind farms totalling just 300MW, largely because of the high cost of erecting and operating turbines at sea. A report commissioned by the Royal Academy of Engineering found that offshore wind was half as expensive again as onshore wind per kilowatt hour, and more than twice as expensive as coal, gas or nuclear.[12] It follows that until the costs come down through economies of scale, offshore wind needs to be given a significant subsidy. The Energy Review proposed to ease the construction of offshore wind farms by streamlining the planning process, and by reforming the existing Renewables Obligation to give more support to technologies that are less economic. This may work, but it is basically tinkering with a system that has already been left for dead by the Europeans. We would have greater certainty of success if we gave it a decent burial and copied their strategy instead.

The scale of the challenge of ramping up non-CO_2 emitting electricity generation quickly to meaningful levels should not be underestimated, especially if the government fails to make progress on any of the areas discussed so far. If we failed to replace our current nuclear capacity, for instance, the amount of electricity produced from wind, say, would have to multiply thirty-seven times before it did anything other than replace the lost nuclear production. We would have to build

9,000 × 3MW wind turbines just to get back to the starting line. We currently have fewer than 1,700 turbines in total.[13] Failure to install CCS would impose a similar handicap. The fact is we need every last scrap of available energy from all these sources simply to stand still, let alone create any surplus to help to mitigate the last oil shock.

Energy efficiency is the mantra favoured by politicians attempting to reconcile the irreconcilable – 'having our planet and eating it'. But a trenchant report from the House of Lords Science and Technology Committee argues that UK policy in this area is profoundly mis-guided.[14] The government places enormous weight on gains in energy efficiency, expecting this to cut carbon emissions by tens of millions of tonnes. It has also claimed that by these means 'across the economy as a whole . . . we could reduce energy use by around 30%'. But the report points out (figure 21) that Britain's energy efficiency has *doubled* since 1970, and *still* total energy consumption has risen.

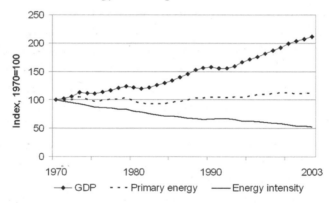

FIGURE 21. UK energy intensity (the amount of energy consumed per unit of GDP) has halved since 1970, meaning its energy efficiency has doubled, but total energy consumption has still risen. Source: House of Lords Science and Technology Committee[15]

This is not a uniquely British but a universal failing, and it is easy to see why it happens. It scarcely matters that modern refrigerators are more efficient than they were, if growing numbers of people now choose models the size of a walk-in closet, along with plasma screen TVs and any number of new electrical gadgets; nor that modern boilers use relatively less gas, if many more families have central heating, and use it to keep a greater number of rooms at a higher temperature; nor that some modern cars and planes do more miles per gallon, if journey numbers and distances continue to explode. In short, left to fight it out, growth trumps efficiency almost every time.

This relationship is no coincidence, and is variously described as the boomerang effect, Jevons' paradox, or the exotic-sounding Khazzoom-Brookes postulate. William Stanley Jevons was a nineteenth-century economist who observed that when James Watt introduced his steam engine, which was far more efficient than Thomas Newcomen's earlier machine, the counter-intuitive effect had been to make total coal consumption rise rather than fall. Jevons concluded that greater efficiency had made coal cheaper to use, and had therefore stimulated demand. This idea – now widely accepted – was further developed by two economists called Daniel Khazzoom and Leonard Brookes, working independently of each other in the late 1970s. It is also perfectly consistent with the work of Professor Robert Ayres, discussed in chapter 5, who argues convincingly that efficiency gains are a fundamental cause of economic growth.

Some of the potential efficiency gains that are now possible are so large that it is tempting to believe that this apparently firm rule will break down in future. For instance, houses built to a German building standard known as *Passivhaus* are so well designed and insulated that they can maintain a temperature of 21°C in the depths of a continental winter without any form of heating – except sunlight and the body heat of their occupants.[16] If Britain legislated and enforced such standards the gas savings would be immense (memo to the minister: *do it*). But the shortcoming of this approach is that however many million energy-efficient homes are built from now on, the preponderance of the existing housing stock is horribly inefficient and also long-lived. The energy efficiency of old buildings can of course be improved, but never to *Passivhaus* standards, and unless demolition rates rise dramatically, 24 million of today's 25 million households will still be standing in 2050.[17]

In another example of the potential of energy efficiency, Woking Borough Council in Surrey has achieved stunning cuts in the energy consumption and carbon dioxide emissions of its buildings largely by converting to Combined Heat and Power (CHP) – small, local gas-fired power stations, which provide both electricity and heat to buildings nearby. CHP is highly efficient because it exploits the large amounts of heat that are normally wasted in electricity generation at large, remote power stations to provide space heating, hot water, and even to drive cooling systems. This doubles the overall efficiency with which the gas is exploited to as much as 90 per cent.

This is naturally a good thing, and it could also be argued that if this kind of efficiency was achieved globally it would liberate gas to be used as a transport fuel and/or delay the gas peak significantly. However there are two important qualifications to make to Woking's undoubted achievement. The first is that although the scheme has doubled its gas efficiency, this has not led gas consumption to halve, because of the general increase in energy demand in the town since the project started in the early 1990s. Woking's Climate Change Project Coordinator, Mick Company, estimates that the scheme has cut actual gas consumption by about 30 per cent. This may be an example of Jevons' paradox at work – or at least starting to.

More importantly, although Woking has cut its gas consumption, its *dependency* upon gas is now even higher than before, in the sense that a far greater proportion of its total energy comes from this single source. Before the scheme started, Woking used to get all its heating and hot water from gas, and all its electricity from the national grid, of which about 40 per cent is supplied by gas-fired power stations. In 2005, having installed CHP along with solar panels and wind turbines, Woking still got all its heating and hot water from gas, but also generated 82 per cent of its own electricity. Of this, however, only 11 per cent came from the renewable sources, and 71 per cent came from the CHP plants, which run entirely on natural gas. So although Woking is now largely protected from the risk of power cuts on the national grid, this comes at the cost of an eye-watering level of exposure to any interruption of gas supplies. London mayor Ken Livingstone has adopted the same CHP approach in order to achieve stiff emissions targets for the capital, but in doing so has also increased the city's vulnerability to a European gas peak in 2010, and a global peak not far behind.

For the time being, Woking and London would be no worse off than anywhere else in the event of a gas supply interruption since they remain connected to the national electricity grid. But if the Woking model were mimicked nationally, the effects could be far more serious. Greenpeace has proposed precisely this strategy – endorsed by Ken Livingstone – in which CHP is expanded aggressively, no new nuclear or even fossil fuel power stations are built, and the national grid is substantially dismantled.[18] If we followed this path and there was a significant interruption to the gas supply – let alone the gas peak – the results could be catastrophic: a total loss of all heat and virtually all power simultaneously. Dismembering the national grid would deny recourse to any form of centralized power – such as nuclear, or even large-scale offshore wind farms – and is therefore a lunatic suggestion.

Despite the impressive emissions reductions and efficiency gains offered by CHP, this can be no more than a short-term strategy, although there is no evidence that its supporters understand this. A document on Woking's website claims that 'greater resistance to supply disruption is provided by meeting local energy needs through locally generated CHP', as if the gas did not come courtesy of Mr Putin, and evidently oblivious of how soon gas may peak.[19] It also claims that when fossil fuels become 'scarce or non-existent', its CHP gas can be replaced by 'biogas, biomass or even hydrogen', although the council has not conducted a resource assessment to justify this assertion. By my back-of-the-envelope calculation Britain could not remotely replace its gas consumption with locally grown biomass once gas peaks, with or without CHP.[20]

It may be that large efficiency gains of the sort achieved by Woking can in fact reduce consumption for a time, but more likely the bigger the efficiency gain, the bigger the eventual boomerang effect. It follows then that efficiency gains cannot be the sole strategy for cutting carbon emissions or mitigating the last oil shock. The House of Lords Science and Technology Committee concluded that the Khazzoom-Brookes postulate 'offers at least a plausible explanation of why in recent years improvements in "energy intensity" at the macro-economic level have stubbornly refused to be translated into reductions in overall energy demand. The government have so far failed to engage with this fundamental issue.' So much for David Cameron's 'green growth' and Gordon Brown's 'new synthesis'. This is not to say that efficiency gains are futile – they are obviously essential – just that they cannot by themselves deliver cuts in overall consumption. Suggesting that they can is like telling the hamster on his wheel that if he runs faster he will eventually get somewhere.

Altogether the three-pronged supply-side policy of nuclear, renewables and efficiency seems unlikely to achieve even the more limited ambition of filling the self-inflicted energy gap and putting greenhouse gas emissions back on track – let alone beginning to deal with the last oil shock. But there is another very simple and effective way to make the numbers stack up: consume less. The question is how to achieve this, since naturally there are few volunteers. And not before time the government is now investigating the introduction of rationing schemes for both companies and people that could – if rigorously executed – achieve just the orderly reduction in fossil fuel consumption that is so urgently needed.

One example of this kind of system already exists in the form of the EU Emissions Trading Scheme (ETS) that covers European electricity generators and other energy-intensive sectors such as steel and cement manufacturing. But unfortunately the early experience of the ETS has mostly served to highlight the weaknesses of a poorly designed system that has been vulnerable to competing national interests.

The scheme is meant to reduce the level of carbon emissions through a 'cap and trade' approach. Each EU government sets an overall ceiling for the emissions of its own power generators and smokestack industries, and shares out the right to pollute proportionately among the relevant companies. They each receive a certain number of permits which each represent the right to emit a tonne of CO_2. Those firms that manage to cut their emissions to below their quota can sell their surplus, while those who emit too much are forced to buy extra allowances. This has created a carbon market which in theory is meant to reward companies that manage to cut and penalize those who fail. The higher the price of carbon units the greater the incentive.

In the spring of 2006 the flaws of the ETS were sharply exposed when the price of carbon units collapsed from more than €30 to less than €10 in a matter of days, as it emerged that nearly all EU governments had succumbed to industry lobbying and issued far too many permits. In 2005 the European industries covered by the scheme emitted 1.785 billion tonnes of carbon dioxide but had been issued with permits to emit 1.829 billion, a surplus of 44 million tonnes which totally undermined the purpose of the ETS.[21] This might have been explained simply as teething troubles were it not for the subsequent behaviour of EU governments. Many countries, including big emitters such as Germany and France, have submitted plans for the second phase of the scheme, which runs from 2008 to 2012, in which their proposed quotas are higher than their existing emissions. Britain, to its credit, has submitted a plan that would impose a cut in its emissions, but the impact of this will be largely neutered if British businesses can simply buy cheap surplus permits from other countries.

The Energy Review declared that British policy was to reform the EU ETS into an effective system with a real scarcity of permits, and to consider introducing a similar UK scheme to regulate large companies such as supermarkets not already covered by the ETS. There was also the briefest reference to the fact that the government was investigating the idea of a similar scheme for individuals, which was later enthusiastically endorsed in a speech by Environment Secretary David Miliband. The minister argued that such a scheme was needed because individuals account for more than 40 per cent of all emissions, and described the idea as 'tradable personal carbon allowances', an ugly but perhaps necessary political euphemism for rationing.

The word rationing immediately conjures up images of wartime austerity, coupons and scissors, and shady black markets (my father still remembers selling his remaining bread ration on the platform at Avignon as he left France in 1949). But a modern system would be quite different: rations would be electronic and effortless to use, they would only relate to fossil fuel consumption, and trading on an eBay-type system would be positively encouraged. Several such schemes have already been worked out, but as it happens I found the simplest and most persuasive had been devised by my neighbour David Fleming, the environmental writer who first alerted me to peak oil. Six years on, it was time to walk up the hill again.

David's scheme of *tradable energy quotas* (TEQs) (formerly *domestic tradable quotas*, DTQs) is described in his lucidly written pamphlet *Energy and the Common Purpose* (www.teqs.net). This is another 'cap and trade' system, but a national one that avoids the obvious shortcomings of the present EU ETS, and which covers all companies and the entire population under one roof. By the time David has finished explaining it all at nineteen to the dozen I find myself oddly enthused about the prospect of living under a system that would force me to make do with less.

Any country introducing the scheme first establishes a carbon budget – how many tonnes of carbon it is going to emit in a single year. The size of the budget is set in advance by an independent Energy Policy Committee – rather like the Bank of England committee that sets interest rates – free from political interference. In the first year the budget will very nearly match the country's actual emissions, but from then on the carbon budget will fall every year. It's a rolling budget, in which the cuts in years 1 to 5 are fixed in advance and unchangeable except by *force majeure*. The cuts planned for the later years can be revised, although there is a strong presumption that they will not be changed. This is a key advantage over the present EU ETS; nobody will be in any doubt about broadly how much less fossil energy they will have to live on in twenty years' time.

The carbon budget is shared out between people, companies and government departments. People are given their rations, but – unlike the EU ETS, and unlike the government's proposed scheme for big British firms – companies and public bodies have to pay for theirs. Every adult is given exactly the same amount, and children are provided for through a system of child allowances. The rations come in units, which could be expressed either as the amount of carbon emitted, or the amount of fossil energy consumed; it amounts to the same thing. Every time you buy petrol, gas or fossil-generated electricity, you will surrender some of your ration through an electronic card or by direct debit. If you consume less than your ration, you can sell the excess through an electronic trading system, something like eBay, and if you consume more, you can buy extra. But availability of additional rations will diminish from year to year as the carbon/energy budget shrinks. If you can't be bothered with the whole thing, you can flog your entire ration as soon as you receive it, but when you buy fuel in future you will automatically be charged for the units necessary to cover your purchase.

The tradable energy quota system is deceptively simple, and the genius of the idea is that it deals with a number of different problems

at once, by creating a series of limits and incentives that once introduced virtually guarantee success: the orderly reduction of fossil
fuel consumption. The system would help to keep us ahead of depletion, making us cut back in a managed way before we are forced
to do so by a potentially violent crisis. 'TEQs are equally designed to
tackle peak oil and climate change, and they involve everybody,' says
Fleming. 'They harness citizen power, and encourage people to think
for themselves.'

The system gives every individual a direct incentive to save energy
far more powerful than those that already exist. The price of the units
will act as both a carrot and a stick: if you exceed your ration you will
have to buy more rations, which could be expensive, but if you reduce
your consumption you will have units to sell, which could be lucrative. Having heard the price of units on the nightly news, you will be
far more inclined to switch off the TV and other appliances at the
socket. The Leeds MP Colin Challen, who is chairman of the Cross-
Party Climate Change Group and an ardent supporter of DTQs, as
they are still more commonly known, argues that such a system could
bring immense domestic energy savings almost instantaneously: 'We
could save ourselves £5 billion within twelve months. There'd be less
fuel poverty and we'd all be better off. That cannot be described as
wearing a hair shirt.'[22]

The system would also give a huge boost to renewable energy production, since electricity from non-CO_2 emitting sources would be
exempt or zero-rated under the scheme. Offshore wind could now be
cheaper than coal- and gas-fired generation, without the need for
subsidies. Customers would switch in droves to renewable energy
suppliers or tariffs, so green companies would win customers and
expand their supply. The system also creates an incentive for radical
efficiency gains throughout the economy – new technologies, and
new ways of doing things – but because overall consumption is
capped, it puts those gains at the service of conservation rather than
greater luxury or higher consumption. This fixes the Khazzoom-
Brookes postulate.

As a policy approach, TEQs are superior not only to the current
EU ETS, but also to the other widely mooted approach – a broad
rebalancing of the fiscal regime away from income tax and towards

'green' taxes. This policy is favoured by the Liberal Democrats, the Conservatives and some elements within the government, but it has several drawbacks, all of which are better solved by a system of tradable energy quotas.

TEQs are fairer than tax reform could ever be. Cutting income tax and raising taxes on energy consumption would inevitably benefit the rich and penalize the poor, who have to spend a far greater proportion of their incomes on energy, even though they generally consume much less. TEQs on the other hand are absolutely fair. 'One of the key features', says Challen, 'is that everybody from the highest to the lowest in the land starts the year with exactly the same quota.' In the early years the rich are able to buy additional rations, but this becomes progressively harder as the carbon budget shrinks. In fact TEQs are more than fair; they are redistributive. In the early years of such a system the poor could benefit by selling their unused rations, and this would particularly help those in 'fuel poverty' – people who spend more than 10 per cent of their income on fuel.

Even if a government did commit to a radical reform of the tax system, however, there could be no certainty that green taxes would ever achieve their intended effect, which is not just to moderate growth, but to achieve large absolute reductions in consumption. But with TEQs the outcome is guaranteed by the existence of the carbon/energy budget. 'The number of units is fixed, so it's impossible to get it wrong,' says Fleming. 'Conformity to the scheme is built into the scheme.' The targets would be achieved regardless of what happened to the price of the units, and probably with much less financial pain than under tax reform.

The impact would not be entirely and brutally mechanistic though. In contrast to tax changes, tradable energy quotas could also transform people's outlook and behaviour rather quickly, according to Matt Prescott, who is conducting a three-year programme to assess the potential of personal carbon trading for the RSA (Royal Society for the Encouragement of Arts, Manufactures and Commerce) in London. 'The real value would be to give people a direct personal stake in the problem and its solution, instead of just leaving it to companies and governments. This is empowering and reduces that sense of helplessness. It could bring about radical changes in consumption patterns.'[23] It might also soften people's resistance to having a wind farm in their back yard.

TEQs would soon become part of the culture. Colin Challen foresees a time when the cost of units will compete with weather and football as an everyday topic of conversation: 'Eventually it will be the thing to talk about in pubs or read about in the *Sun* – what you're doing, how much carbon you're saving, how much money you're saving as a result, and what you're going to spend it on.' Since other people's behaviour will affect the price of the units, excessive energy consumption will become as socially unacceptable as drink driving, the SUV as unfashionable as the shell suit.

As the carbon/energy budget shrinks, people will be encouraged to think up new and co-operative ways of cutting their consumption, such as car-sharing. To arrange trips so that the maximum number of people travel in a single vehicle would cut the cost in units for everybody, and also reduce the number of car journeys, fuel consumption and emissions. The system would soon start to get people out of their cars altogether, tipping the balance in favour of walking, cycling or public transport for shorter or discretionary journeys.

TEQs would also transform the behaviour of companies because – unlike existing cap and trade schemes – they will be forced to pay for all their units. Under the EU ETS companies in most countries are given their entire initial entitlement free of charge, and only shell out if they exceed their emissions quota and have to buy more permits in the market. And under the government's proposed scheme for big British companies, according to the Energy Review, 'auction revenues would be recycled to participants'. This dulls the incentive to change the way they do business. Under TEQs, however, companies have a much greater spur to invest in clean power generation, and to reduce fossil fuel consumption in every aspect of their business. Companies' costs will rise and so may the price of their goods and services, particularly those that are fossil fuel intensive such as air fares. But competition will mitigate price rises, because energy and emissions reduction will become a key competitive advantage: cleaner firms will be able to undercut their dirtier rivals. The system provides an imperative for companies to source supplies locally, and this will throw up new opportunities for entrepreneurs and domestic agriculture. Supermarkets will have to rethink and localize their entire supply chains.

The other great benefit for companies is the stable planning horizon: knowing for the first time ever exactly what will be required of them twenty years hence. This will stimulate investment in research and development to produce the energy efficiency and other break-throughs that will be required to satisfy the shrinking carbon budget of future years. As a result countries that jump first will sacrifice nothing, but rather gain a valuable head start. 'Any country doing this early would be at a substantial and increasing competitive advantage,' argues Fleming, 'because its energy costs would go down and because its companies would have to produce lots of energy-efficient tech-nologies which could then be sold all over the world.' But don't take his word for it: in June 2006 thirteen major companies including Shell, Vodafone and Tesco trooped in to see Tony Blair to ask him to impose the toughest possible CO_2 targets for just this reason.[24]

The fact that jumping first confers competitive advantage offers some grounds for optimism that such a scheme could spread around the world relatively quickly, without the need for fractious and long-drawn-out negotiations, and the laggards will be the losers. The sight of other countries making off with tomorrow's industries and profits might just drag the United States into the twenty-first century, and lead it to choose survival over a temporary extension of American hegemony – otherwise known as going down with all guns blazing. Despite Bush's summary rejection of Kyoto, dozens of American cities have already signed up voluntarily to cut their emissions, while a group of seven states in the north-eastern US is developing a carbon trading scheme similar to the EU ETS, California has passed a law imposing industrial CO_2 reductions, and peak oil activism is every-where on the Internet. This is all evidence of a burgeoning popular movement that could yet develop into an effective political force for change.[25]

Although TEQs do not depend on global treaties to make them work, this rationing system – or something like it – will form an essen-tial part of any overarching agreement to tackle either climate change or the last oil shock. Each country will find its own way to satisfy inter-national obligations, but TEQs would be an effective and flexible method to achieve a number of different goals. The Rimini Protocol proposed by Colin Campbell, founder of the Association for the Study of Peak Oil, is the first draft of a simple treaty that is intended to minimize conflict and keep prices manageable as oil production decline sets in.[26] Oil exporters would promise to run down their reserves no faster than at present, while importers would commit to cutting their consumption to match the global decline rate. Prices would then moderate, poor countries could still afford to import, and if everybody stuck to it there would be nothing to fight about. Every country would be forced to take responsibility for cutting its own consumption – probably using TEQs – but could at least be sure that as the pie shrank, its slice would stay proportionally the same.

One disadvantage of the idea is that existing inequalities would be locked in. 'Of course it's not fair that the US should continue to consume so much,' admits Campbell, 'but you have to be realistic and give them time to adjust.'[27] But even if the US agreed to such a scheme – get real – the rest of the world might justifiably baulk at this concession to continued American over-consumption.

Another approach that is attracting support among policymakers internationally is known as *contract and converge,* which although designed principally to tackle climate change could also be usefully deployed against oil depletion. Contract and converge is based on the idea that the only safe and fair way to apportion the right to pollute the world can be on an equal per capita basis.[28] A global agreement would establish scientifically the upper limit of atmospheric carbon concentration that can be tolerated, divide the 'acceptable' emissions between the world's population, and set a date by which all countries have to hit this per capita level. After a time, each country's cap would be fixed regardless of its subsequent population growth. This overall approach implies that some developing nations such as China could continue to increase their emissions for a time, while the industrialized nations would have to begin steep cuts immediately, again probably using TEQs. The biggest polluters such as America would obviously have to cut hardest, but even countries such as Britain would have to make enormous reductions. If the agreement were to stabilize atmospheric CO_2 at 550 parts per million – which many scientists regard as dangerously high – UK emissions would have to fall from the current ten tonnes per person per year to less than four tonnes by 2050.[29] Four tonnes is the amount generated by one person taking a round trip by plane between London and New York. For this reason, despite the undoubted fairness of the idea, it would be just as hard to secure agreement around contract and converge as the Rimini Protocol.

Although TEQs would offer competitive advantage to big business, selling the idea to a sceptical public will not be easy. People generally do not understand the urgency of the threats, many, including the elderly, might be bewildered by the trading scheme, and there would no doubt be vociferous opposition to anything that raised the price of travel, especially air fares. 'You can just imagine the headlines,' says Matt Prescott, whose programme of work for the RSA is intended to iron out many of the remaining practical and policy issues raised by personal carbon rationing, and crucially to solve the problem of how to win public acceptance. 'How to change attitudes is a really tough nut to crack,' says Prescott, who plans to launch a pilot scheme by 2009, something like an airmiles programme in reverse, which awards points to people for reducing their consumption or emissions. The idea would be to get people used to the idea of carbon as a currency that can be traded, and to start with the scheme would of course be voluntary. If the plan and its timetable seem to lack urgency, Prescott argues the politics must proceed step by step: 'If we came out tomorrow in favour of rationing, the tabloids would have a field day and it would be a public relations disaster. But at the same time people are saying this is so urgent we really need to be doing it now. Both things are true, but unless we get this right we'll be shot down in flames and then we really are in trouble. We've probably only got one chance to get this right.'

Whether New Labour or any other party has the political will to see it through, however, is quite another matter. On the evidence of the Stern Review and the propaganda barrage that accompanied its publication in October 2006, the answer is a resounding no. To hear Blair and Brown you might have thought that not only the government had all the answers, but also that none of them would hurt. The chancellor boomed that Stern meant Britain could be 'pro-growth and pro-green', while the Prime Minister declared 'you don't have to restrict growth or your lifestyle'.[30] This is indeed the message of the Stern Review, but unfortunately it is based on a fundamental fallacy.

Stern offered us the choice of paying 1 per cent of GDP now in order to prevent the impact of climate change crushing the economy by as much as 20 per cent eventually, but the review failed to acknowledge that this apparent bargain offer is likely to be swept away by the last oil shock. In 600 pages Stern made no mention of the oil peak, and dismissed the significance of resource depletion in a single paragraph, buried deep in the report: 'There is enough fossil fuel in the ground to meet world consumption demand [sic] at reasonable cost until at least 2050.'[31] But as we saw in chapter 8 the oil peak will probably precipitate the very economic crisis Stern ascribes to climate change – which be likened to the Depression and the Second World War rolled into one – within the next decade or so. Who then would or could pay the 1 per cent – if indeed the real cost is so little – necessary to mitigate global warming? As I have already argued, the last oil shock will not help but hobble our attempts to save the planet.

The launch of the Stern Review was accompanied by fine words from Gordon Brown about new targets, a big rise in the energy research and development budget, international agreements and a global carbon market, but his speech was also laden with the usual delusional statements about biofuels, hydrogen, technology and efficiency gains. It was also notable for the things it failed to mention.

Despite the letter from Environment Secretary David Miliband, leaked on the eve of the publication of the Stern Review, urging Brown to impose a range of stiff 'green taxes', the chancellor scarcely mentioned the subject, and his record shows no real appetite for such a strategy. The government's craven policies on air travel, petrol duty and SUVs to date suggest that any changes in this area will continue to be timid.

Neither did Brown mention personal carbon trading; the carbon market he eulogised was only for British companies and other countries, not domestic voters. But perhaps this was all a political softening up exercise, designed to prepare the ground for the introduction of personal carbon trading sometime soon. I hope so, but from the language I doubt it. The longer New Labour and their opponents remain paralysed by political cowardice and fail to prepare the country for the last oil shock, the worse the eventual crisis will be.

In the end the politicians' fear may be swept away by events. David Fleming is convinced that the early tremors of the last oil shock will force the government to introduce his TEQs scheme, or something very similar: 'I insist that it will happen. Having exhausted all other possibilities the government will eventually do the right thing. We'll be very lucky if we get through the next few years without a serious outage of oil or gas, and then the government will have no option.'

TEQs, or something like them, will be a vital part of any response to the last oil shock, but they are not a panacea. Our quick review of the shortcomings of British policy suggests a range of further steps that must also be taken:

1. Scrap the DTI. As someone once said of another department, it is 'not fit for purpose'. Replace it with a Department for Energy and Climate Change, with a cabinet-ranking Energy Secretary. This is not just a name change, but year zero for a brand-new department. It follows therefore that senior civil servants would have to reapply for 'their' jobs.
2. The new department's mission statement is no longer 'creating the conditions for business success', but *achieving complete independence from hydrocarbons by 2030, by expanding renewable supply where environmentally acceptable and managing demand as necessary*.
3. Energy policy to be informed not only by the current price of commodities, but also strategic resource assessments. Achieving energy security now means picking the obvious winners and rigging markets aggressively to support them.
4. Send the new Energy Secretary on a fact-finding mission to Sweden, Denmark, Germany and Spain to see how they get things done. Come home, legislate and enforce.
5. Commit to one more generation of nuclear, but also match that quickly with at least the same amount of offshore wind. This is used to displace half our gas-fired generation, which may liberate gas for use in transport. If it doesn't create such a surplus, then by definition we need to be cutting gas use anyway. Maintain the existing coal capacity, but install CCS immediately for the twenty worst polluters. Taken together, this provides at least our current level

of electricity supply for another generation, with much reduced emissions, and may also mitigate the oil peak.

6. Having delivered point 5, keep on expanding offshore wind, subsidizing if necessary. It's clean and secure and there's more than enough. A couple of additional pence per kilowatt hour is absurdly cheap. Our eventual goal is to source all our heat, power and transport fuel from renewable sources, and offshore wind could provide a huge proportion.

7. Massively increase the UK energy research and development budget. The government has moved in this direction by setting up the new National Institute for Energy Technologies, but more could be done. Intensive research should be directed to achieving breakthroughs in electricity storage technologies; adaptation of the national grid to accommodate a far higher penetration of wind turbines and other intermittent power generation; and the development of biofuel crops suitable for sub-agricultural land.

8. Launch an urgent international, scientific assessment of the potential for biofuels production in the tropics – what, where and how much – and devise aid and trade incentives to stimulate a rapid growth in production, along with strict governance mechanisms to ensure that this does not threaten food production, nor critical ecosystems such as the rainforests. Environmental groups could play a monitoring role.

9. A fundamental overhaul of the teaching of economics at university, founded on the proposition that there is no economy without a planet, explicitly incorporating the laws of thermodynamics, and giving far greater prominence to the work of non-economists such as Kümmel, Hall and Ayres.

10. Start the diplomacy to achieve an international settlement on the lines of the Rimini Protocol, and/or contract and converge.

None of this programme is especially radical, except perhaps for point 2. But even achieving complete independence from hydrocarbons by 2030 should be possible, provided policymakers accept that energy consumption must fall massively, and that to achieve this markets must be taken by the scruff of the neck. Extraordinary things can be achieved when society is put on a wartime footing, which would be entirely appropriate to our situation. All it needs is some brave political leadership. What a terrifying thought.

II

How to Survive the Imminent Extinction
of Petroleum Man

I F POLITICAL LEADERS cannot be relied upon to do the right thing
in time, what can we do in our own lives to prepare for the last oil
shock? If you broadly accept the analysis I have offered so far, the
general implications for our personal conduct should be fairly clear.
In many areas the responses required will be the same as those urged
on us by climate change campaigners – only more so and with even
greater urgency. The immediate motivation is rather different – not
altruism but self-preservation – although in practical terms it will
amount to the same thing.

Three principles stand out immediately. First, complete insulation
from the last oil shock is impossible, but the lower your oil and gas
dependency the better protected you will be. Second, adjusting to a
lower energy lifestyle will take time, so those who start now will be
better prepared when the crisis breaks. We should all start to cut our
fuel consumption immediately and progressively, and not wait for the
introduction of carbon or energy rationing. And third, this is not 'just'
about transport, because our ability to travel long distances at whim
has shaped almost every aspect of how we live today. Once you start
tugging at this thread, all sorts of areas start to unravel, so coping with
the last oil shock will mean re-engineering our lives from top to
bottom. Let's start with transport.

There are all sorts of ways for motorists to cut their emissions and
fuel bills right away, although most of them offer only limited protection against the last oil shock. The first thing to do is reassess your
driving behaviour. Whatever the car, you can save fuel by following a
few simple rules: drive slowly and gently, because accelerating hard
and slamming on the brakes is thirsty work; don't idle unnecessarily,
so kill the engine whenever you can; keep the tyres pumped up; and

use the air conditioning sparingly, since it increases fuel consumption by up to 14 per cent, but remember that keeping the windows open at speed creates so much drag it wastes even more energy.[1] In short, drive like an old man not a boy racer. Jot down your mileage on the back of the receipt each time you buy fuel to keep tabs on your miles per gallon. If you stick to it, all this will improve your fuel economy, but depending on how well you drive already, the difference could be marginal.

For many people, however, it should be possible to achieve a radical improvement in miles per gallon simply by changing car. If for some strange reason you happen to own an SUV, the best advice is to flog it; what do you suppose will happen to the second-hand value of your 20mpg beast when oil hits $200 a barrel? Owning one is not only selfish but stupid. And even if your current car is not obscenely wasteful, it is still possible to improve your fuel efficiency by an astonishing margin, particularly if you are prepared to trade down in size. The Toyota Prius hybrid four-door saloon gets about 55mpg, the two-seater petrol-driven Smart Fortwo does almost 60mpg, while the CDI turbodiesel version (left-hand drive, available in Europe only) gets about 80mpg.[2] Three other models – Peugeot 107, Citroën C1 and the Toyota Aygo – which are all made in the same factory in the Czech Republic, do almost 69mpg.[3]

There are also cost and emissions savings to be made by changing the type of fuel your car runs on. Most petrol-driven cars can easily be converted to run on other hydrocarbons, the most common alternative being LPG (liquefied petroleum gas, made of propane and butane), available at over 1,200 filling stations in Britain (www.boostlpg.co.uk). It might cost £2,000 to convert the car to run on both fuels, but the subsequent savings in the UK could be as much as 40 per cent. For the same price you can also convert to CNG – compressed natural gas – and even install a compressor to refuel from your domestic gas connection at home (www.gasfill.com), although this would cost another £2,000, and the taxman would expect to hear about it.

LPG and CNG are certainly much cheaper than petrol and diesel at the moment, but this advantage may not survive the last oil shock. Most of the cost advantage is due to lower fuel duty because the government wants to encourage use of these cleaner alternatives, so as the take-up rises, the tax advantage will probably be withdrawn. Then the price of the fuel will be more closely linked to the underlying gas price, which as we have seen moves in line with oil. Globally, natural gas will peak later than oil under any scenario, although regional gas peaks could come sooner rather than later, and with brutal effect. In the short to medium term supply disruptions are equally likely with either fuel.

Electric vehicles are also perhaps better suited to tackle climate change than the last oil shock. In Britain, the tiny G-Wiz electric car (www.goingreen.co.uk) claims to do the equivalent of 600mpg, at a fuel cost of around 1p per mile, as against about ten times as much for a mid-sized petrol-engine car.[4] They are also exempt from road tax and London's congestion charge and parking fees – so for a daily commuter they could pay for themselves from day one. And provided you also change to a green electricity provider such as Good Energy (www.goodenergy.co.uk), which generates all its power from wind and other renewables, in theory you could continue to enjoy carbon-and-guilt-free city motoring that also seems amazingly cheap.

The G-Wiz is however something of a niche product. Some of the shine is taken off its impressive fuel economy by the need to replace the battery every three years at a cost of £1,200 – a problem suffered by all electric cars – and the financial advantages are far lower if you don't live in London and commute into the West End on a daily basis. In fact, according to Dr Ben Lane, an environmental consultant and author of the *Green Car Buyers' Guide* (www.ecolane.co.uk), in most scenarios for people living outside London and driving less than about 10,000 miles per year, electric cars work out more expensive per mile than conventional. 'You still have an environmental advantage but not necessarily a financial advantage.'[5] And as we saw in chapter 8, the price of electricity is likely to soar in line with fossil fuels, so owning an electric vehicle will not necessarily insulate you from the short-term impacts of the last oil shock.

But electric vehicles have a potentially significant role in the longer term if battery technology continues to improve, as it almost certainly will, and if Britain can overcome its self-inflicted electricity deficit. If hydrogen cars ever make it into mass production, electric cars will still have one overwhelming advantage: because they refuel by plugging directly into the grid, they suffer none of the huge upstream energy losses involved in the production and transport of hydrogen (chapter 4), and this means the number of wind turbines needed to power an equivalent number of vehicles would be far smaller.

Nevertheless, electric cars are still handicapped by their small size, low range (the latest G-Wiz claims up to forty-eight miles on a single charge, if you don't turn the heater on), and the fact that you need a garage or at least a driveway at home in order to be able to recharge the vehicle from a domestic electricity socket, because draping the cable over the pavement is illegal. All these problems are soluble, but will take time to solve.

Although any of these cars or alternative fuels could help to cut your emissions and your fuel bill – and so give some financial cushion against the last oil shock – the statistical likelihood is that having bought a car that is more efficient, you will simply use it to drive further and negate much of the benefit. This is the Khazzoom-Brookes postulate or 'boomerang effect' at a personal level. Even if you managed to resist this temptation and kept your mileage the same, you would not have reduced your *dependency*, which should be gauged not simply by the amount of money you spend on fuel, but rather by how much your life would be disrupted if you had to do without it. Clearly, in a world of carbon rationing and fuel shortages, the fewer miles you need to drive the less vulnerable your lifestyle. Cutting mileage is the only reliable way to minimize your exposure to the last oil shock, and since it will take time to adjust, the time to start is now.

Colin Challen started by trading in his old Ford Escort for a Smart Fortwo, halving his engine size, increasing his fuel economy, and cutting his emissions by 30 per cent at a stroke. Then he went on to reduce his annual mileage from 8,000 to 6,000, and plans to halve it eventually. As a busy constituency MP he achieved this by combining greater use of public transport and a Brompton folding bicycle, which he loves (I went for a Dahon, but there are lots to choose from). 'It doesn't mean wearing a hair shirt,' says Challen, 'but it does mean changing to other forms of transport in a managed, thoughtful way.' Buying a bicycle and using it regularly is probably the easiest and most effective way of making that change.

If the idea of halving your mileage sounds daunting, you don't have to achieve such a large cut immediately, although to minimize your exposure to the last oil shock you do need to get working on it. You also need to keep in mind that the final goal may well turn out to be eliminating personal car use altogether, and to be aware of where you are starting from. The average British car clocks up just over 9,000 miles a year, and the average main driver in a household does 7,400.[6] If your totals are higher than this, cutting your mileage is even more urgent.

One approach could be to set yourself an initial target to halve your mileage over five years at 10 per cent a year, but monitor your progress weekly. Start by checking your annual mileage from your servicing records or last two MOT certificates, divide it by fifty-two, knock off 10 per cent, and then see which of your regular, weekly trips could achieve the reduction. Is there one kind of car journey you could cut back or eliminate entirely – the school run, trips to the supermarket, or commuting? If the early cuts are easy, think about what further changes you would need to make to repeat the reduction the following year, and the year after that. You may find quite quickly that you cannot achieve even your first five-year plan without starting to make fundamental changes to many aspects of your daily routine.

The whole exercise is obviously far easier if your home is within walking or cycling distance of school, shops and work. If it is, you may already be in a position to ditch your car altogether, perhaps joining a car pool (www.car-pool.co.uk) for the occasions when you still need to use one. If you truly cannot reach any of these essential locations except by car, you probably need to start thinking about moving home, or lobbying your local authority for some decent public transport. If neither is an immediate prospect, whenever you do next move house, proximity to decent local amenities should be at the top of your tick list, along with the energy efficiency of your new home.

In Europe there are probably relatively few homes that are quite so car dependent, but in America the entire landscape of strip malls and suburbs will be largely unliveable in after the oil peak, the impact of which is explored in a powerful documentary available on the Internet (www.endofsuburbia.com). It is the fact that the American suburbs – and the American dream – need to be grubbed up and rebuilt to cope with the last oil shock that makes US foreign policy reaction to oil shortage so dangerous. There are some signs of progress, however, in the growth of the New Urbanism movement, with planners and architects beginning to build small towns in which a mixture of housing, shops, businesses and recreation are all located within walking distance, and where the design encourages mass transit over car use (www.calthorpe.com, www.dpz.com). But the challenge of converting the existing sprawl into mixed and sustainable communities is immense. In the meanwhile, if you find yourself living somewhere without even a corner shop for miles around, you may be sitting on a business opportunity.

Depending on where you live, the way you shop can help to reduce your exposure to the last oil shock – especially how you choose to use supermarkets. Although they now seem indispensable, supermarkets are a problem because their very existence encourages car use. If you can do all your shopping locally on foot or by public transport, so much the better, but if not, it should at least be possible to cut the frequency of supermarket trips. If your budget and storage space can cope, you could organize a big shop for staple goods once a month, perhaps using the supermarket delivery service, and in between times buy all your perishable food locally on foot.

What you buy also matters, wherever you buy it. We saw in chapter 5 the astonishing distances that some produce is transported, and how a basket of supermarket items might have travelled six times round the planet. After the oil peak, these products will either become far more expensive or disappear altogether, so it is worth getting used to eating domestically produced fruit and vegetables in season, and other products that are at least from the same continent. As a start I recently banished New World wine from my shopping cart – no great sacrifice with France and Italy next door – along with antipodean meat and butter. Buying British is a good policy, but in supermarkets this is no guarantee that the produce is not well travelled; some British-grown apples are sent to South Africa to be waxed before being returned for sale in Britain.[7] (This, apparently, is the market 'working'.) The one way you can be sure is to shop at a farmers' market, or join a box scheme where locally produced fruit and veg are delivered to your home or to a nearby collection point every week (www.vegboxschemes.co.uk).

Of course the way to cut your food miles to an absolute minimum, if you have a garden or an allotment (www.allotment.org.uk), is to grow your own. Doing so also allows you to trim some energy consumption by composting all your organic household waste (www.greencone.com). This not only displaces fertilizer, which as we saw in chapter 5 is highly dependent on natural gas, but also cuts the fuel consumption of the bin men's lorry – or would if everybody did the same. And that's quite apart from the delicious food you will produce for little more than the cost of a packet of seeds (www.theboxingclevercookbook.co.uk).

At home, the aim should be to minimize your consumption of heating oil and natural gas, and to try to eliminate their use altogether if possible. This is far easier if you can afford to invest in a range of electricity-generating and alternative-heating technologies, but if not you can still achieve substantial savings. Again, the first thing to change is your own behaviour. In winter, wearing warmer clothes and turning the thermostat down by a degree or two could save lots of fuel – about 10 per cent for every 1°C reduction on average.[8] You should also consider whether you need to heat all of your home to the same temperature all the time, turn off every last appliance includ-ing videos and set-top boxes when not in use, and resolve not to dawdle in the shower: I'm down to under two minutes and nobody has yet complained.

The next step is to make some small but cost-effective investments in the energy efficiency of your house. Make sure the insulation for your roof, cavity walls and hot water tank is up to modern standards – it's cheap to replace and will quickly pay for itself. Double glazing is expensive, but refurbishing and draught-proofing your windows is much cheaper and very effective. Fit energy-saving light bulbs throughout as the old ones pop. Information on all this, and a range of grants for which you may be eligible, is available from the Energy Saving Trust (www.est.org.uk).

It should be possible to dispense with oil and gas entirely for home heating, provided you have a garden and some money to invest, using a device known as a ground source heat pump (www.ukheatpumpnet.org). The machine is about the size of a large fridge, and works just like a fridge in reverse, by circulating fluid through a loop of pipe buried under the lawn, which absorbs heat and transfers it to a hot water tank, even in the depths of winter. The system is best suited to underfloor heating systems – so you might have to dig up the house as well as the garden – but can provide hot water too. It can also work backwards, providing cooling in the summer. Electricity is needed to pump the fluid round, but the system typically produces between two and a half and four times as much energy as it consumes. They are not cheap, but once installed should last for some thirty years.

Provided you have also switched to a wholly renewable electricity supplier, with a ground source heat pump your heating and hot water would be emissions-free, but you would still be exposed to the last oil shock through the price of electricity. You could reduce this by installing a wind turbine, and if you installed two might just about cover the entire electricity requirement of the ground source heat pump for an average house, provided the turbines perform at their rated capacity. But do your homework, because experts at the Energy Saving Trust are concerned that turbines may produce less than a quarter of their official output when sited in urban environments, where buildings can reduce the average windspeed substantially.[9]

There are three British domestic wind turbine designs on the market: Swift (www.renewabledevices.com) is more powerful, Windsave (www.windsave.com) is cheaper, and Stealthgen from Eclectic Energy (www.stealthgen.co.uk) the most unobtrusive. Installation is fairly straightforward, and the Windsave model simply plugs straight into a domestic 13amp socket. There are still some drawbacks however. Vibrations caused by turbines could damage the brickwork of older buildings, so take advice. The power companies have not yet come up with a simple system to pay householders for excess electricity that they export back to the grid, although the government has set the industry a deadline for the end of 2007. And by law all domestic wind turbines are designed to stop generating during power cuts – just when you might have thought you needed them most – to protect electricians working on the network. So if you want a measure of protection against blackouts you will have to consider some kind of separately wired system with batteries.

Photovoltaic solar panels – the kind that generate electricity – are still by far the most expensive form of domestic power generation per kilowatt hour, and could take decades to pay for themselves at current prices. Having said that, electricity prices are likely to keep rising, which can only make all forms of home generation – and ground source heat pumps – more attractive. So if you are moving house or doing major renovation work, it might make sense to incorporate some PV panels, which are now available as bolt-on systems or integrated roofing tiles (www.currys.co.uk, www.solarcentury.com).

However the other kind of solar panel, which heats water directly, is probably the most cost-effective way to cut your gas bill after effective insulation. A roof-top solar thermal system costing £2,500–£4,000 could provide a family of four with as much as half of its hot water over the course of the year – although it is obviously most productive during the summer – and for the next twenty years at least (www.nef.org.uk). The water is pumped through tubes on the roof to be warmed by the sun, and then to the hot water tank, where the temperature can be topped up by a conventional boiler when necessary. Such a system can also be made to work in combination with a ground source heat pump, with both feeding the same tank. Solar hot water needs electricity to drive its own pump, but some systems are cleverly designed to be totally self-sufficient by using a solar PV panel to provide the power (www.solartwin.com).

However much you invest in electricity generation at home you are very unlikely to meet all your needs all the time. That means that until the entire electricity grid is free of fossil fuel generation, and for as long as you continue to consume natural gas at home, you remain exposed to the last oil shock. As a result it makes sense to keep trying to cut your overall consumption as much as possible. Again, be methodical: monitor your gas and electricity consumption, and keep a record. Dig out your old bills and compare them with the previous year, bearing in mind the seasonal variations. If your gas meter measures in cubic metres, convert the reading to kWh by multiplying by 10.8 to compare your gas and electric consumption on a like-for-like basis.[10] All this will help you to work out which of your ideas achieve the greatest energy reductions.

I also recommend you buy a plug-in electricity meter (www.brennenstuhl.com, www.doctorenergy.co.uk), to help to work out which devices use most power. You will probably be as shocked as I was to find out quite how much your gadgets waste. Even when my little Roberts radio is switched off, the mains transformer that powers it draws just under 10 watts. If I left it like that for a year, I would have wasted enough electricity to keep a one-bar electric fire blazing for three and a half days, or to drive a G-Wiz electric car almost 230 miles – London to Liverpool.[11] On further investigation I find a handful of these greedy little black boxes plugged in around the flat, each consuming about the same amount of power. But the worst culprits are the TV, video and cable box, which together draw over 30 watts on standby, and the computer and all its peripherals, which draw well over 50 watts – *even when everything appears to be fully switched off, never mind on standby*. Left for a year, the TV, computer and their associated gadgets would waste enough power to drive the G Wiz over 2,000 miles, or London to Athens.[12] Needless to say, I now switch everything off at the mains religiously. It's easier to keep up the habit if you ensure that as many related gizmos as possible are fed from one, easily accessible socket. Alternatively, you could buy an intelligent mains adaptor that automatically cuts the power to the accessory devices when the television or computer is turned off or on standby (www.oneclickpower.co.uk).

Some modern gadgets are energy savers however. If you cook with electricity, your new toy will tell you that the microwave consumes far less power than the traditional electric oven, so it would pay to use it as much as possible. But beware what you put in it – ready meals contain a lot of embodied energy – and remember to turn the microwave off at the socket when you have finished; the digital clock also wastes power, and for most people serves no purpose. Whether you cook with electricity or gas, a few simple rules such as boiling only as much water as you really need, and keeping the lid on, will also cut your energy consumption noticeably.

The wider economic impacts of the last oil shock such as inflation and unemployment will be even harder to avoid than the direct effects on energy supplies, but there are ways to reduce your exposure. We've seen already that it makes sense to work close to home, but the kind of job you do is also important, and some sectors are obviously more vulnerable than others. The prospects for travel agents and airline staff will nosedive with the last oil shock, but on the other hand there will be huge pressure on governments to find ways to fuel and expand public transport. The outlook for car workers also looks dire, but at least their engineering skills should be transferable to the production of wind turbines, wave generators and so forth. Retraining and shifting careers before the crisis hits would be a smart move.

Ironically the oil and gas industry will probably continue to hire, whatever the vicissitudes of the oil price and share values. For one thing, as the average size of oil and gas fields falls, oil companies will need relatively more staff to produce each barrel. For another, the same companies are sure to dominate the development of carbon capture and storage, and by some estimates stuffing the CO_2 genie back in the bottle could become as big a business as releasing it ever was. And finally, to survive in the longer term oil companies will be forced to segue into the new energy economy, a move that some are already beginning to make.

As a result the traditional disciplines of geology and engineering will offer good employment opportunities, as will any trades that help to create the new energy infrastructure, including construction, electrical engineering and agronomy. The Internet will become even more important because of the increasing constraints on travel and as teleworking finally comes into its own. Accountancy, strangely, will also expand, since companies will need a new specialist breed of energy-literate bean counters to conduct carbon audits and administer TEQs. If you have any influence over your children's educational and career choices, you could increase their resilience to the last oil shock by nudging them in the direction of any of the above, encouraging them to take science over liberal arts, and, please, anything but media studies.

The impact of the last oil shock on economies and stock markets is likely to be profound, bringing into question much of the conventional wisdom about personal investment. While it is impossible to predict exactly what will happen, in the worst-case scenario markets could collapse and remain permanently depressed, making pensions and endowment policies close to worthless. At the same time, inflation could rise to levels not seen for generations. Exposure to these aspects of the crisis is inescapable, so the question is how to minimize the impact. I should stress that I am not qualified to offer investment advice, but then few of those who are so qualified know anything about the last oil shock, and most have a direct financial interest in encouraging you to put your money into the stock market where it is now at greater risk. It is important to calibrate their advice against what you now know about the impending crisis.

The first point to make is that spending now to reduce your fossil fuel dependency is in effect a risk-free pension investment. If you intend to stay in your present home, investing in energy efficiency and micro-generation will allow you to live in greater comfort at lower cost during your retirement, whatever your income. So although I would never recommend anybody *not* to invest in a pension, despite the increased market risks, if you are weighing up whether to top up your pension or invest in some of the ideas discussed above, remember that a stock market crash can never take away your solar hot water system.

A stock market crash could, however, take away your job, and indeed your home if you can't keep up with the mortgage payments. So the other imperative is to pay off your debts as quickly as possible. I realize that this advice conflicts with the need to spend money to reduce your fossil fuel dependency, but that is in the nature of the dilemma. Consider scaling back holidays and other discretionary spending to achieve as much of the strategy as possible. The alternative is to borrow to the hilt and pray for hyperinflation to get you out of trouble, but in those circumstances you could soon be out of work and lose everything.

If you are fortunate enough to have invested in reducing your energy dependency, paid off your debts, and *still* have money to spare, I am full of envy. You could now invest in oil and gas stocks, which will be lucrative in the short term, although as the oil peak approaches, this will only be for the investor who enjoys the most violent fairground rides. Or you could back a range of renewable energy companies, which may not be quite so racy in the short term, but where demand for their products in the longer term is more or less guaranteed (www.renewableenergystocks.com).

As with any self-help programme, there is of course a long list of things to give up, or at least cut back on. What follows is a handful of the most obvious. **Plastic shopping bags**: we get through

17.5 billion a year in the UK, 100 billion in the US, and up to 1 trillion globally.[13] Think ahead, take your own bag or rucksack, and save an oil well. **Bottled water**: an absurd marketing-driven fad in any country where the tap water is potable. The world consumed 153 billion litres in 2003, wasting energy not only to make the glass and plastic bottles, but also to cart the weight of all that liquid from country to country. This is one area where Western European per capita consumption apparently exceeds that in the US – 112 litres vs 90.[14] Buy a camping bottle and fill it at the tap, and if you don't like the taste use a filter. **Non-seasonal fruit and veg**: if your 'organic' produce has come from halfway round the world, the label is meaningless greenwash and your purchase is both wasting oil and needlessly raising CO_2 emissions. **Patio heaters**: just a wicked waste of natural gas. **Meat:** may become prohibitively expensive as grain prices continue to soar. About a third of global grain production goes to feed livestock, but grain prices are already rising sharply as supplies come under added pressure from drought, failing crops, population growth and the increasing proportion of US maize being diverted to produce ethanol for road fuels (20 per cent and rising in 2005).[15] World grain production has failed to match consumption in six of the last seven years, and stocks are dwindling.[16] So rebalance your diet towards Mediterranean rather than American quantities of meat, and dust off that vegetarian cookbook you have ignored for so long. **SUVs**: because they directly threaten your Sunday roast, and will soon be killing the world's hungry.

But it's not all about abstinence and giving up; think of all the ways your life will improve as you adapt to the last oil shock! Robert Hirsch and his colleagues who wrote a report about how to mitigate the effects of the oil peak for the US Department of Energy could only come up with one positive impact: less junk mail. But they can't have been trying very hard, because I can think of loads. **Get fit, lose weight**: courtesy of your new bicycle. **Save money**: immediately on gym subscriptions and road fuel, and eventually on road tax, MOT, insurance and every other cost associated with car ownership. **Get there faster**: the average traffic speed in London is 10mph, slower than a century ago, and if you can't top that on a pushbike then you probably need the exercise anyway.[17] **Make money**: by living within your energy budget and selling some TEQs. **De-stress**: no more fuming in interminable jams, as traffic volumes fall and the roads clear for improved public transport. **Protect your children**: who will be healthier and safer, particularly if you live anywhere near busy roads, not to mention having a greater chance of inheriting a habitable planet. **Holidays closer to home**: Britain has fabulous landscapes and beaches, and don't worry – it will get warmer. **Satisfaction**: that you did what you could. Others may continue to squander energy, and the crisis will not be solved by your voluntary abstinence – that's why carbon or energy rationing is essential – but whatever the outcome, you will be better off than if you had done nothing.

This is not remotely to suggest, however, that the overall impact of the last oil shock will somehow turn out to be benign. The odds are certainly stacked against such a happy outcome, not least because the oil peak is likely to converge with rapid climate change and a crisis of global agriculture, and because each will exacerbate the others. As Lester Brown of the Earth Policy Institute argues forcefully but soberly in *Plan B: Rescuing a Planet under Stress and a Civilization in Trouble*, the world's fish stocks are collapsing, grain harvests since the turn of the century have consistently failed to match consumption, and aquifers everywhere are being over-pumped, virtually guaranteeing a sharp drop in food production at some point in the not too distant future. Meanwhile the world's population heads towards 9 billion by the middle of the century.

Climate change will worsen the food crisis, as rising temperatures destroy the glaciers that feed the world's great rivers, and suppress photosynthesis in the principal crops (harvests slumped dramatically around the world simply as a result of the heatwave of 2006). Some of the world's most productive agricultural land is likely to be inundated by rising sea levels, cutting output and creating tens of millions of refugees. The oil peak will probably worsen both climate change *and* the food crisis. Dwindling oil supplies will encourage the maximum exploitation of coal and oil sands, generating far more CO_2 than conventional crude, although the additional liquids will fail to fill the supply gap. As that gap widens, yet more food crops will be diverted into fuel production, putting the gas tanks of the rich in direct competition with the stomachs of the poor. And finally fuel shortage will threaten agricultural output directly, since every calorie of food you consume takes ten calories of fossil fuel to produce.

In fact we seem to be shaping up for the very combination of crises first explored by the environmental thinktank the Club of Rome in their famous book *The Limits to Growth* back in the 1970s. Economists the world over will tell you this work has been 'disproved', but most of them clearly haven't even bothered to read it. At least, that's the charitable interpretation. For the record, the principal prediction contained in this book has not yet fallen due: it *cannot* have been disproved.

The central thesis of *The Limits to Growth* was that continued exponential growth of population and consumption in a finite world would eventually lead to an apocalyptic global crisis and mass starvation. The limits to growth would come either in the form of resource constraints – food, fuel, metals, water – or pollution, or both. Remember, this work was published long before climate change and the oil peak were widely recognized threats. Yet now we face just such a series of interlocking crises, of which it seems the last oil shock will be only the first instalment. And thirty years on, the timing of the Club of Rome's overall prediction looks ominously on track. In 1972 they wrote: 'If the present growth trends in world population, industrialization, pollution, food production and resource depletion continue unchanged, the limits to growth on this planet will be reached sometime within the next one hundred years. The most probable result will be a rather sudden and uncontrollable decline in both population and industrial capacity.'

If *The Limits to Growth* is finally to be proved wrong, humanity must now quickly solve a cluster of further crises in the midst of the likely financial, economic and military spasms of the last oil shock. And it must achieve this while many of the most important opinion-formers – oilmen, environmentalists and politicians – persist in denying or ignoring what is by now obvious. Their behaviour makes it more likely that we will hit the wall at full speed, rather than braking hard and choosing a safer route. Yet I have to believe that the outcome is still within our control – if only just. Welcome to the last oil shock. Thanks for reading. Good luck.

Notes

Preface

1. Jean Laherrère, interviewed by the author, August 2006; Colin Campbell, interviewed by the author, July 2006; Richard Hardman, interviewed by the author for *The Money Programme: The Last Oil Shock*, BBC2, on 11 November 2000.

Chapter 1. Sources in Washington

1. *Petroleum Review*, January 2000, carried the text of Cheney's speech as written.
2. Verbatim transcript of Cheney's speech as delivered, provided by the *Petroleum Review*.
3. Craig Unger, *House of Bush, House of Saud*, Gibson Square Books, 2004.
4. Peter Galbraith, interviewed by the author, May 2005.
5. For some reason it is often claimed that Iraq has the world's second largest reserves, but this is wrong, at least according to the most commonly quoted industry sources. Figures from both the *BP Statistical Review* and the *Oil and Gas Journal* give Iran's proved reserves as bigger than Iraq's, and their relative status remains the same even if you add in their respective 'yet-to-find' figures calculated by the USGS.
6. Kenneth Katzman, *The Iran-Libya Sanctions Act (ILSA)*, Congressional Research Service, Report for Congress, updated 31 July 2003.
7. US Energy Information Administration (EIA), *Global Energy Sanctions*, July 2004, www.eia.doe.gov/emeu/cabs/sanction.html.
8. Cheney's speech, provided by the *Petroleum Review*.
9. The Royal Bank of Scotland *Oil and Gas Index* (www.rbs.com) gives the UK peak as 2.7 mb/d in April 1999, excluding natural gas liquids. The *BP Statistical Review*, which includes NGLs as well as crude oil, gives the peak as 2.9 mb/d during 1999, but does not specify the month (www.bp.com).

10. UK Offshore Operators Association, www.ukooa.co.uk.

11. Department of Trade and Industry, *The Oil and Gas Industry Task Force Report: A Template for Change*, September 1999, Annexe A, 'Vision Workgroup: Economic Advisory Group Report'.

12. Ibid.

13. Average North Sea production was 2.909 mb/d in 1999 including natural gas liquids, and 2.667 mb/d in 2000, according to the *BP Statistical Review*, www.bp.com.

14. USGS, *World Petroleum Assessment 2000 – Description and Results*, USGS Digital Data Series DDS-60 (CD-ROM). Proved reserves from the *BP Statistical Review*, www.bp.com.

15. Average monthly production January 1997 to December 1999 was just over 1.9 mb/d, according to both the *Oil and Gas Journal*, and *BP Statistical Review*.

16. John Pilger, *The New Rulers of the World*, Verso, 2002.

17. Security Council Resolution 661, 1990, cited in Professor Joy Gordon, *The Machinery of Annihilation: The US and the Iraq Sanctions, 1990–2003*, Harvard University Press, manuscript in preparation.

18. *Report of the second panel established pursuant to the note by the president of the Security Council of 30 January 1999 (S/1999/100), concerning the current humanitarian situation in Iraq*, Annex II of S/1999/356, 30 March 1999, para. 11, cited in Professor Joy Gordon, *Machinery of Annihilation*.

19. UNICEF and the Government of Iraq, *Situation Analysis of Children and Women in Iraq*, April 1998, cited in Professor Joy Gordon, *Machinery of Annihilation*.

20. Carne Ross, interviewed by the author, April 2005.

21. Thomas J. Nagy, 'The Secret Behind the Sanctions: How the U.S. Intentionally Destroyed Iraq's Water Supply', http://www.progressive. org/, and *Washington Post*, 23 June 1999.

22. UNICEF and the Government of Iraq, *Child and Maternal Mortality Survey, 1999*, cited in Pilger, *New Rulers of the World*.

23. Cited in Pilger, *New Rulers of the World*, Albright claims in her memoir *Madam Secretary* that she did not mean to say this, and that Saddam could have prevented the deaths had he done what he was told. In other words, America was happy to use the suffering of innocents as a tool of foreign policy. Cf http://www.fff.org/comment/com0311c.asp.

24. Denis Halliday wrote in the *Seattle-Post Intelligencer*, 12 February 1999: 'Even the most conservative, independent estimates hold economic sanctions responsible for a public health catastrophe of epic proportions. The World

Health Organization believes at least 5,000 children under the age of 5 die each month from lack of access to food, medicine and clean water. Malnutrition, disease, poverty and premature death now ravage a once relatively prosperous society whose public health system was the envy of the Middle East. I went to Iraq in September 1997 to oversee the U.N.'s "oil for food" program. I quickly realized that this humanitarian program was a Band-Aid for a U.N. sanctions regime that was quite literally killing people. Feeling the moral credibility of the U.N. was being undermined, and not wishing to be complicit in what I felt was a criminal violation of human rights, I resigned after 13 months.'

25. Personal communication, United Kingdom Offshore Operators' Association, 1 June 2005. The comparison seemed reasonable since both Iraqi and UK North Sea production peaked at about 3 mb/d, although in different years.

26. *Report of the group of experts established pursuant to paragraph 12 of Security Council resolution 1153 (1998)*, http://www.un.org/Depts/oip/background/reportsindex.html.

27. *Report of the group of United Nations experts established pursuant to paragraph 30 of the Security Council resolution 1284 (2000)*, http:// www.un.org/Depts/oip/background/reportsindex.html.

28. *Report of the United Nations team of experts established pursuant to paragraph 15 of Security Council resolution 1330 (2000)*, published May 2001.

29. www.simmonsco-intl.com.

30. *The Money Programme: The War for Oil*, BBC2, 26 March 2003.

31. 'Warning Shot that Spurred Blair', *Guardian*, 13 September 2000.

32. Personal communication, Cabinet Office official, 5 September 2005.

33. 'Poll-Axed', *News of the World*, 17 September 2000.

34. The oil industry gave $2.5 million to the Bush campaign and almost $27 million to the Republicans in total. Source: http://www.opensecrets.org.

35. Analysts believe Chevron's oil output will rise in 2006 and beyond, following its acquisition of Unocal in late 2005, and other new projects it has in development. However, the company's oil production fell from just over 2 mb/d in 1998 to 1.7 mb/d in 2005 (source: PFC Energy). In February 2005 Dave O'Reilly, Chevron's chairman and CEO, became one of the first oil company bosses to talk publicly about depletion. In a speech in Houston he said: 'relative to demand, oil is no longer in plentiful supply. The time when we could count on cheap oil and even cheaper natural gas is clearly ending,' and concluded: 'we are seeing the beginnings of a bidding war for Mideast supplies between East and West.' Cf http://www.chevron.com/news/speeches/2005/.

36. The former Director of Intelligence at the State Department Carl W. Ford Jr, giving evidence to the Senate Foreign Relations Committee, described Bolton as a bully, a 'serial abuser' of junior staff, and 'a quintessential kiss-up, kick-down sort of guy'.

37. PNAC, *Rebuilding America's Defenses: Strategy, Forces and Resources for a New Century*, September 2000, http://www. newamericancentury.org.

38. Also to be found at http://www.newamericancentury.org.

39. PNAC, *Rebuilding America's Defenses*.

40. 'Bush Energy Paper Followed Industry Push', *New York Times*, 27 March 2002.

41. Jane Mayer, 'Contract Sport: What did the Vice-President do for Halliburton?', *New Yorker*, 16 March 2004.

42. *National Energy Policy: Report of the National Energy Policy Development Group*, May 2001, www.whitehouse.gov/energy/ National-Energy-Policy.pdf.

43. Jim Munns, interviewed by the author, April 2005.

44. DTI, *Our Energy Future – Creating a Low Carbon Economy*, February 2003, http://www.dti.gov.uk/energy/.

45. Quote from Christopher Meyer, *DC Confidential*, Weidenfeld & Nicolson, 2005; 'Two Minds on the Middle East: Despite the Gung-ho Headlines, an Uneasy Blair is Trying to Moderate Bush's Unilateralism', *Guardian*, 8 April 2002.

46. 'Iraq action is delayed but "certain"', *The Times*, 8 April 2002.

47. Ibid. The article was clearly well sourced: 'Military action aimed at toppling Saddam Hussein is likely to be delayed for at least a year but is certain to happen because it is an "article of faith" for President Bush, senior American and British diplomats have disclosed . . . During prolonged talks at Mr Bush's Prairie Chapel ranch at the weekend, Tony Blair and the President are believed to have envisaged a three-phase plan to tackle the Iraqi dictator: first recreating a coalition for action by convincing Iraq's neighbours that Saddam can be ousted; then taking military action against him; and finally ensuring that a successor regime is capable of running the country.'

48. Letter from White House Chief of Staff Andy Card to Commerce Secretary Donald Evans dated 20 May 2002: 'During his recent stay at Crawford, Prime Minister Blair suggested, and the President agreed, that we should have a bilateral US-UK energy dialogue. Blair's Chief of Staff Jonathan Powell advises that Secretary of State for Trade and Industry, Patricia Hewitt, is their choice to lead the dialogue on the UK side.' The 'energy security and diversity' quote comes from a Memorandum for the President, 30 July 2003, and accompanying report, first released to the *Guardian*, which can be

found at http://image.guardian.co.uk/sys-files/Guardian/documents/
2003/11/14/abraham_docALL.pdf.

49. 'UK and US in Joint Effort to Secure African Oil', *Guardian*, 14 November
2003, http://www.guardian.co.uk/guardianpolitics/story/
0,,1084903,00. html.

50. Letter from Patricia Hewitt to Don Evans, dated 11 July 2002, released to
the *Guardian* by the US Commerce Department. I am grateful to Rob
Evans for sending me a copy.

51. Personal communication, Foreign and Commonwealth Office, 4 and 19
April 2005.

52. Personal communication, Foreign and Commonwealth Office, 11 February
2005.

53. Evans's and Hencke's documents came from the US Department of
Commerce. Mine came the Department of Energy and the State
Department, as well as Commerce.

54. *Nigeria Focus*, July and August 2002, www.menas.co.uk.

55. Jonathan Bearman, interviewed by the author, April 2005.

56. This is from an undated document headed 'US-UK Energy Dialogue'
released to the *Guardian* by the US Commerce Department. Thanks again
to Rob Evans.

57. This is from a longer version of the progress report written for Blair and
Bush, but which is not available on the *Guardian* website. Yet again, thanks
to Rob Evans.

58. Carne Ross, interviewed by the author, April 2005.

59. *The Money Programme: The War for Oil*.

60. The despatch was dated 2 September 2002, and headed 'Scenesetter: U.S. –
UK Energy dialogue, September 9th', in preparation for a meeting of senior
officials in Washington.

61. Deutsche Bank, 'Baghdad Bazaar', 21 October 2002; Centre for Strategic
and International Studies, 'Iraqi Oil . . . The Morning After', 9 January
2003.

62. A number of smaller international oil companies have started to operate in
Kurdistan where the security situation is much calmer and the regional government is keen to establish its primacy over Baghdad in relation to oil
resources in the region. But, according to the Middle East Economic Survey
(Vol XLIX, No 36, 4 September 2006), so far, the larger IOCs have not been
prepared to risk offending Baghdad by dealing with the KRG.

63. Richard Perle said on 11 July 2002: 'Support for Saddam, including within
his military organisation, will collapse at the first whiff of gunpowder'; Ken
Adelman, former US Ambassador to the UN, on 13 February 2003: 'I believe

demolishing Hussein's military power and liberating Iraq would be a cake-walk'; and Dick Cheney on 16 March 2003: 'I really do believe we will be greeted as liberators.' Source: *Observer*, 30 March 2003.

64. 'Return to the Fold: How Gaddafi was Persuaded to Give up his Nuclear Goals', *Financial Times*, 27 January 2004.
65. Personal communication, USGS official, 1 June 2005.
66. Seymour Hersh, 'The Coming Wars: What the Pentagon Can Now Do in Secret', *New Yorker*, 24 January 2005.
67. Robin Cook wrote: 'I put these points to the prime minister a couple of weeks later. The exchange is recorded in my diary on March 5 2003. Tony Blair gave me the same reply as John Scarlett, that the battlefield weapons had been disassembled and stored separately. I was therefore mystified a year later to hear him say he had never understood that the intelligence agencies did not believe Saddam had long-range weapons of mass destruction.' 'Blair and Scarlett Told Me Iraq Had no Usable Weapons', *Guardian*, 12 July 2004.

Chapter 2. Dangerous Curves

1. *The Tuesday Documentary: The Energy Crunch: 1*, BBC, 12 June 1973.
2. Ronald Doel, Dept of History, Oregon State University, writing in the *Handbook of Texas Online*, http://www.tsha.utexas.edu/handbook/online/articles/view/HH/fhu85.html.
3. Kenneth Deffeyes, *Beyond Oil: The View from Hubbert's Peak*, Hill and Wang, 2005.
4. All original Hubbert quotes are taken from the oral history interviews of the American Institute of Physics (AIP), conducted shortly before Hubbert's death in 1989. The full transcripts are available at http://www.hubbertpeak.com/hubbert/aip/.
5. M. King Hubbert, 'Nuclear Energy and the Fossil Fuels', API, 1956. You can see the entire paper at http://www.hubbertpeak.com/hubbert/1956/1956.pdf.
6. Dennis Meadows, Jorgen Randers, Donella Meadows, *Limits to Growth: The Thirty-Year Update*, Chelsea Green, 2004.
7. Source: M. King Hubbert, 'Nuclear Energy and the Fossil Fuels', 1956, reproduction courtesy of the American Petroleum Institute.
8. Ibid.
9. Source: Roger Bentley, 'Global Oil and Gas Depletion: An Overview', *Energy Policy*, 30, (2002), 189–205, reproduction with permission from Elsevier.

10. Personal communication from an official at the UK Foreign Office, 11 February 2005.

11. Charles A. S. Hall, Cutler J. Cleveland and Robert Kaufmann, *The Ecology of the Economic Process: Energy and Resource Quality*, John Wiley and Sons, 1986.

12. It is important to grasp that this graph is distinctly different from the graphs Hubbert drew in 1956. Those showed the *rate* of production: any point on the graph line tells you the amount of oil produced in a single year. This graph is *cumulative*: any point on the graph line tells you the total oil discovered or produced up to that moment. Once no more oil is being produced the cumulative graph line will flatten to horizontal, but it can never fall back towards the bottom axis; that would mean oil was being undiscovered, or unburned and stuffed back into wells. Source: *Energy Resources. A Report to the Committee on Natural Resources of the National Academy of Sciences-National Research Council*, Publication 1000-D, Washington, National Academy of Sciences-National Research Council, 1962. Available at: http://www.hubbertpeak.com/hubbert/EnergyResources.pdf. Reproduction courtesy of the US National Academy of Sciences.

13. Ibid.

14. M.K. Hubbert, *Oil and Gas Supply Modelling, Proceedings of a Symposium held at the Department of Commerce, Washington DC, June 1980*, US Dept of Commerce, May 1982. This is Hubbert's swansong paper and reviews all his methods. I recommend it.

15. A.D. Zapp, 'Future Petroleum Producing Capacity of the United States', *Geological Survey Bulletin*, 1142-H, 1962.

16. M.K. Hubbert, 'Degree of Advancement of Petroleum Exploration in the United States', *American Association of Petroleum Geologists*, Bulletin 51, November 1967.

17. This graph first appeared in the 1967 paper (ibid.), but without the curved line that makes the exponential decline of the bar chart so much clearer. I have taken this reproduction from Hubbert, *Oil and Gas Supply Modelling*, simply because the line has been added.

18. Kenneth Deffeyes, *Hubbert's Peak: The Impending World Oil Shortage*, Princeton University Press, 2001.

19. Daniel Yergin, *The Prize: The Epic Quest for Oil Money and Power*, Simon & Schuster, 1991, gives a detailed account of the first oil shock. Astonishingly, however, in this 900-page, thoroughly researched tome, Mr Yergin could not find space for a single reference to M. King Hubbert.

20. IHS Energy.

21. Stewart Udall, interviewed by the author, June 2005.

22. Hall, Cleveland and Kaufmann, *The Ecology of the Economic Process*.

Chapter 3. The Wrong Kind of Shortage

1. The embargo against Iraq after its rejection of Security Council resolution 1352 in 2001 took more than 2 mb/d off the market; the strike at Venezuelan state oil company PDVSA at the end of 2002 shut in up to 2.6 mb/d for three months; and the invasion of Iraq again took out as much as 2.3 mb/d for the best part of 2003. Source: IEA, *Fact Sheet on IEA Oil Stocks and Emergency Response Potential*, 2005, www.iea.org.

2. 'To Conserve Gas, President Calls for Less Driving', *New York Times*, 27 September 2005.

3. IEA, *Oil Market Report*, 9 September 2005, www.iea.org.

4. *BP Statistical Review*, June 2005; IEA, *Oil Market Report*, 9 September 2005 and 11 October 2005, www.iea.org.

5. The total discovery from three years is summed and divided by three, and attributed to the central year of the three. For example, discovery for 1964, 1965 and 1966 is added together, divided by three and attributed to 1965.

6. Graham Dore, interviewed by the author, August 2005.

7. IHS Energy.

8. *Shell Global Scenarios to 2025*, 2005.

9. IHS Energy.

10. *The Money Programme: The Last Oil Shock*, BBC2, 11 November 2000.

11. Dr Campbell, interviewed by the author, February 2005.

12. For America the R/P rose from about 10 in 1980 to 11 in 2004, although production has fallen 40 per cent below its peak, and Indonesia's R/P has risen since 1991 even though its production has halved.

13. IHS Energy actually calls its series 'resources', but it is their equivalent of proved and probable.

14. Roger Bentley, interviewed by the author, October 2005.

15. Ken Chew, *World Petroleum Trends, 1994 to 2003*, October 2004, personal communication. Ken Chew's analysis suggested real reserve growth over the course of the decade of about 175 billion barrels. The reason this growth is not evident in figure 13 is that for technical reasons this kind of reserve growth is backdated to the date of the original oil discovery. This means the proved and probable line will tend to maintain its downward slope, but float a bit higher each year. However, in 2004 and 2005, discovery and reserve growth failed to match oil consumption for the first time.

16. *The Money Programme: The Last Oil Shock*.

17. Ibid.

18. Harry Longwell, 'The Future of the Oil and Gas Industry: Past Approaches, New Challenges', *World Energy*, Vol. 5, No. 3, 2002, http://www.worldenergysource.com/articles/pdf/longwell_WE_v5n3.pdf.

19. Hubbert's global forecast also missed because of the delayed impact of the first two oil shocks which depressed global consumption and therefore production. Until about 1980 the forecast was on track.

20. USGS, *World Petroleum Assessment 2000*.

21. T. R. Klett, Donald L. Gautier and Thomas S. Ahlbrandt, 'An Evaluation of the U.S. Geological Survey World Petroleum Assessment 2000', *AAPG Bulletin*, August 2005.

22. Jason Nunn, interviewed by the author, January 2005.

23. Chris Skrebowski, 'The Winners and Losers in 2005 Production', *Petroleum Review*, July 2006.

24. The full list, by year of peak production, comes courtesy of Dr Michael Smith of Energyfiles.com: Poland (1909), Czech Republic (1953), Austria (1955), Kyrgyzstan (1961), Germany (1968), Bulgaria (1969), Libya, Ukraine, United States (all 1970), Israel (1971), Kuwait (1972), Turkmenistan (1973), Belarus, Iran (both 1974), Romania (1976), Indonesia (1977), Tajikistan, Trinidad and Tobago (1978), Ghana, Morocco, Myanmar (1979), Peru (1980), Chile, Croatia (1981), Albania, Georgia, Hungary (1982), Spain, Tunisia (1983), Barbados, Cameroon, Greece (1984), Benin (1985), Netherlands (1986), Congo (Kinshasa), Egypt, Russia, Taiwan (1987), France, Senegal (1988), Pakistan, Turkey (1991), Japan (1992), Papua New Guinea (1993), Serbia and Montenegro (1994), Syria (1995), Slovakia (1996), Gabon, New Zealand (1997), Argentina, Uzbekistan (1998), Colombia, United Kingdom (1999), Australia (2000), Bahrain, Norway, Oman, Surinam (2001), Cuba, Yemen (2002), Denmark, Mexico, South Africa (2004) and Italy (2005). Oil production in a handful of these countries – Russia, Libya, Iran and Peru – is actually rising, but not expected to exceed their previously established peaks. The rest are in outright decline.

25. 'ExxonMobil Senior Vice President Discusses New Resources and Advanced Technology', *Business Wire*, 11 January 2005. The report noted that Exxon's Senior Vice-President Stuart McGill 'noted that world demand for oil and gas is expected to increase by 1.7 percent per year, while the world's oil and gas fields on average are declining in production at a rate of 4 to 6 percent per year. This base decline, coupled with the growing demand for oil and gas, means that the amount of new daily production needed in 2020 is nearly equivalent to replacing all of today's daily production.'

Chapter 4. Long Fuse, Short Fuse

1. The Kyoto Treaty obliges about 40 industrial countries to cut their collective greenhouse gas emissions to 5.2 % below 1990 levels by 2012, but the agreement to cut emissions does not include the world's biggest polluter, the United States, or countries in the developing world such as China and India whose emissions are growing fast.
2. 'Kyoto Protests Disrupt Oil Trading', *Guardian*, 17 February 2005.
3. Floor trading was finally abandoned on 7 April 2005; now all trades are done electronically.
4. www.earth-policy.org; 'Sea Level Rise Doubles in 150 Years', *Guardian*, 25 November 2005; DTI Energy White Paper, *Our Energy Future – Creating a Low Carbon Economy*, February 2003.
5. IPCC, *Third Assessment Report*, September 2001, www.ipcc.ch.
6. 'Warming Hits "Tipping Point"', *Guardian*, 11 August 2005.
7. Professor Paul Wignall, interviewed by the author, November 2005.
8. Global emissions in 2004 were just over 26 billion tonnes, and the IEA Reference Case forecast for 2030 is 40.4 billion tonnes. If oil peaks in 2010 and output declines at 3 per cent a year, global emissions would be 32.2 billion tonnes. *World Energy Outlook* © OECD/IEA. 2006, Figure 2.8, p. 81, as modified by the author.
9. IEA, *World Energy Outlook 2004*, www.iea.org.
10. 'Schwarzenegger Unveils Hydrogen Hummer, But Not How He Planned', Associated Press, 23 October 2004.
11. 'Arnold's Hydrogen Hummer', *Forbes*, 4 January 2005, www. forbes.com/vehicles/2005/01/04/cx_dl_0104vow.html.
12. 'Fuel Cells: Japan's Carmakers are Flooring It', *BusinessWeek* online, 23 December 2002, www.businessweek.com.
13. NFPA, 1991, cited in J. L. Alcock, *Compilation of Existing Safety Data on Hydrogen and Comparative Fuels*, May 2001, http://www.eihp.org/ public/documents/ CompilationExistingSafetyData_on_H2_and_ComparativeFuels_S..pdf.
14. Scientists are working on other forms of storing hydrogen, such as tanks filled with metal hydrides or other hydrogen-absorbent materials, which amazingly manage to hold more gas than the empty volume of the tank, without being pressurized (don't ask me how). But these technologies rely on containers that are either extremely heavy or require temperatures of about $-200°C$, and all are a long way from being commercially useful. In 2004 Professor Mark Thomas of the Northern Carbon Research Laboratories at Newcastle University made a significant breakthrough by developing a lighter, synthetic

'molecular sponge' in which to store hydrogen, but he says the technology is probably 'decades' from commercialization. For him storage is the biggest technical obstacle to widespread use of hydrogen. 'It may not be solved,' he told me. 'I don't think there's any guarantee.'

15. Unless otherwise sourced, these numbers were provided by BOC, a company that has been supplying hydrogen to the chemicals and petroleum industry for 130 years.

16. Amory Lovins, 'Twenty Hydrogen Myths', June 2003, www.rmi.org. Lovins, a huge proponent of hydrogen, cites 9–12 per cent.

17. Malcolm A. Weiss *et al.*, *Comparative Assessment of Fuel Cell Cars*, Massachusetts Institute of Technology, February 2003, http://lfee.mit.edu/public/LFEE_2003-001_RP.pdf. NB This study takes into consideration not only the energy used to produce the various fuels and run the cars, but also the energy used to make the vehicles themselves.

18. 'Chemicals Group Exposes Itself to the Whims of the Spot Market', *Financial Times*, 2 August 2005.

19. [a] UK mileage figure from National Statistics, www.statistics.gov.uk; US figure from Bureau of Transportation Statistics, www.bts.gov. [b] Split the difference between mileage per kilogram of hydrogen of the fuel cell Ford Focus and the Honda FCX 2005, www.corporate.honda.com. [c] 25kWh consumed in the process, 40kWh left in the resulting kilo of hydrogen. BOC, personal communication. [d] In Britain passenger vehicles account for two-thirds of total fuel consumption, freight one-third (source: DTI, *UK Energy Sector Indicators 2005*, www.dti.gov.uk). To estimate the total including freight therefore means increasing the figure by 50 per cent. [e] The figure of 4 turbines per km^2 was provided by Peter Hinson, whose company Your Energy Ltd builds wind farms. [f] According to the British Photovoltaic Association 10 square metres of standard photovoltaic panels has capacity of 1kW peak (www.greenenergy.org.uk). Therefore 1 square kilometre has a peak capacity of 100 MW. [g] If a 1kW solar array could produce at peak capacity all the time, it would produce 8,760 kWh hours a year. But since it's dark at night and can be cloudy during the day, it doesn't. Average output achieved in Britain is 750kWh (www.greenenergy.org.uk), giving a load factor of just 8.56 per cent. To compensate for this, the size of the array needs to be increased by a factor of 11.9 (100 / 8.56 = 11.9). Because it has more sunshine America probably achieves 65 per cent greater output on average than Britain (Whitfield Solar, personal communication), so in the US example I have given a load factor of 14 percent, meaning theoretical capacity must be multiplied by 7. [h] UK nuclear capacity 11,852MW (www.dti.gov.uk); US nuclear capacity about 100,000MW (source: EIA, US Department of Energy, www.eia.doe.gov).

20. Britain's commercial power generators had capacity of 73,308 megawatts at the end of December 2004; source: http://www.dti.gov.uk/energy/inform/energy_stats/electricity/index.shtml.

21. This sounds odd, since oil contains more carbon than gas does, but results from the inefficiency of fossil fuel power generation. Source: Nick Eyre, Malcolm Fergusson, Richard Mill, *Fuelling Road Transport, Implications for Energy Policy*, Department for Transport, November 2002.

22. State of the Union address, 28 January 2003, www.whitehouse.gov. Bush presented hydrogen *explicitly* as an alternative to regulating the auto industry: 'In this century, the greatest environmental progress will come about not through endless lawsuits or command-and-control regulations, but through technology and innovation. Tonight I'm proposing $1.2 billion in research funding so that America can lead the world in developing clean, hydrogen-powered automobiles. A single chemical reaction between hydrogen and oxygen generates energy, which can be used to power a car – producing only water, not exhaust fumes. With a new national commitment, our scientists and engineers will overcome obstacles to taking these cars from laboratory to showroom, so that the first car driven by a child born today could be powered by hydrogen, and pollution-free. Join me in this important innovation to make our air significantly cleaner, and our country much less dependent on foreign sources of energy.' All the facts I have reported were well known by the time this speech was written.

23. Personal communication, Dr Jeremy Woods, Imperial College, London; DTI, Dukes 3.4–3.6, www.dti.gov.uk/energy/statistics/source/oil/page18470.html.

24. IEA, *Biofuels for Transport – An International Perspective*, 2004, www.iea.org. The East of England Development Agency and the National Farmers' Union maintain that Britain can meet its biofuel target by using land that has been 'set aside' under the Common Agricultural Policy, and by 'diverting' wheat exports. But if every country did the same, somebody would end up hungry. Source: East of England Development Agency, *The Impacts of Creating a Domestic UK Bioethanol Industry*, 2003.

25. Some authorities such as Professor David Pimentel argue that ethanol production in America actually consumes more energy than it produces. Even ethanol's supporters only claim it delivers 1.3 to 1.8 times the fossil energy that goes into producing it. Source: Ethanol Across America, *Net Energy Balance of Ethanol Production*, autumn 2004.

26. 'It's Corn vs. Soybeans in a Biofuels Debate', *New York Times*, 13 July 2006.

27. Lester Brown, 'Ethanol Could Leave the World Hungry', *Fortune*, 21 August 2006.

28. Susan M. Wood and David B. Layzell, *Canadian Biomas Inventory: Feedstocks for a Bio-based Economy*, BIOCAP Canada Foundation, 27 June 2003, prepared for Industry Canada, http://www.biocap.ca/ index.cfm?meds=section&section=36&category=20. Canada's gasoline consumption was 40.3 billion litres in 2004, http://www.ctv.ca/servlet/ ArticleNews/story/CTVNews/20050928/gas_consumption_050928?s_na me=&no_ads.

29. Worldwatch Institute, *The State of the World 2006*, chapter 4, www.worldwatch.int.

30. Ibid. Presentation by consultant Alfred Szwarc, Brazilian Ministry of Science and Technology, 'Use of Bio Fuels in Brazil', 9 December 2004, http://unfccc.int.

31. [a]Ministry of Agriculture, Livestock and Food Supply, *Brazil as a Strategic Supplier of Fuel Ethanol*, January 2005, www.ethanol-gec.org. [b]Ibid. [c]UNICA, *Prospects for the Sugar and Ethanol Industry*, November 2005, www.brazil.org.uk. [d]A gallon of petrol contains 116,090 British thermal units of energy, whereas a gallon of ethanol contains 76,330 BTUs. Source: Ethanol Across America, *Net Energy Balance of Ethanol Production*, autumn 2004, www.ethanolacrossamerica.net. [e]IEA, www.iea.org/textbase/stats/ oilresult.as, figures given in tonnes, converted at 8.5 barrels of gasoline per metric tonne, cf. *BP Statistical Review 2005*.

32. Professor Jim Ratter, interviewed by the author, December 2005.

33. Satellite surveys suggest that 66 per cent of the Cerrado has already been converted to human use, and that the remainder is disappearing at a rate of 1.1 per cent per year. Source: R. B. Marchado *et al.*, 'Estimativas de perda da area do cerrado brasileiro', unpublished technical report, Conservacao Internacional, www.conservation.org.br. At this rate it would take about thirty years.

34. Dr Paulo Wrobel, interviewed by the author, December 2005.

35. UN Food and Agriculture Organization, www.fao.org.

36. IEA, 'World Energy Outlook', 2006.

37. World Wildlife Fund, 'China Funds Massive Palm Oil Plantation in Rainforest of Borneo', 12 August 2005, http://news.mongabay.com/2005/0812-wwf. html; 'Palm Plantation Sparks Rainforest Row', CNN, 16 September 2005, http://edition.cnn.com/2005/WORLD/asiapcf/09/16/indonesia.palm/.

38. 'Feeding Cars, Not People', *Guardian*, 23 November 2004.

39. [a] IEA, www.iea.org/textbase/stats/oilresult.asp; [b] D1 Oils.

40. Jet kerosene consumption in 2003 was 204,937,000 tonnes, or 4.4 mb/d. Source: IEA.

41. Robert Saynor, Ausilio Bauen and Matthew Leach, *The Potential for Renewable Energy Sources in Aviation*, 7 August 2003, www.iccept.ic.ac.uk.

Chapter 5. Last Oil Shock, First Principles

1. According to ExxonMobil the figure was 7.7 mb/d in 2003, or 9.8 per cent of that year's supply. The company forecasts the chemical industry will consume 11.4 mb/d in 2020, or 10.8 per cent of global supply. Personal communication.
2. American Chemistry Council, *Guide to the Business of Chemistry*, 2004, www.americanchemistry.com; 'The End of Cheap Oil', *National Geographic*, June 2004; Norman Baker MP, Liberal Democrat Shadow Environment Secretary, *How Green is Your Supermarket? A Guide for Best Practice*, March 2004; http://www.reusablebags.com/facts.php; International Federation of the Phonographic Industry, www.ifpi.org.
3. [a] NatureWorks says that operating at full capacity its plant would consume 13 million bushels of corn (39 bushels per tonne) to produce 140,000 tonnes of PLA. www.natureworks.com. [b] American Plastics Council. [c] US Census Bureau, *Statistical Abstract of the United States: 2004–2005*. [d] Ibid.
4. 'Ethanol's Demand for Corn May Trim US Meat Output', Reuters, 23 May 2006, www.planetark.com/dailynewsstory.cfm/newsid/36482/story.htm.
5. Daniel Yergin, *The Prize,* Simon & Schuster, 1991.
6. Andrew Oswald, 'Oil and the real economy', 17 March 2000; 'Can the "New Economy" really survive expensive oil', January 2000. http://www2. warwick.ac.uk/fac/soc/economics/staff/faculty/oswald/.
7. Alan A. Carruth, Mark A. Hooker and Andrew J. Oswald, 'Unemployment Equilibria and Input Prices: Theory and Evidence from the United States', *Review of Economics and Statistics*, 1998. Reproduction with permission from MIT Press.
8. Robert Solow, 'A Contribution to the Theory of Economic Growth', *Quarterly Journal of Economics*, February 1956.
9. Charles Hall, Dietmar Lindenberger, Reiner Kümmel, Tim Kroeger and Wolfgang Eichhorn, 'The Need to Reintegrate the Natural Sciences with Economics', *BioScience*, August 2001.
10. Professor Charles Hall, interviewed by the author, June 2005.
11. Robert U. Ayres and Benjamin Warr, 'Accounting for Growth: The Role of Physical Work, *Structural Change and Economic Dynamics*, February 2004.

12. Amory Lovins, 'How America Can Free Itself of Oil – Profitably', *Fortune*, 4 October 2004. 'When the U.S. last paid attention to oil efficiency, between 1977 and 1985, oil use fell 17% while GDP grew 27%. During those eight years, oil imports fell 50% and imports from the Persian Gulf fell by 87%.'

13. Robert L. Hirsch, Roger Bezdek and Robert Wendling, 'Peaking of World Oil Production: Impacts, Mitigation and Risk Management', February 2005, which can be found at the website of the US DoE's National Energy Technology Laboratory, www.netl.doe.gov.

14. Robert Ayres, interviewed by the author, June 2005.

15. IXIS Corporate & Investment Bank, 'The Price of Oil in 10 Years' Time: USD 380 a Barrel', 18 April 2005.

16. US Bureau of Transportation Statistics, *2001 National Household Travel Survey*.

17. Oakridge National Laboratory, *Transportation Energy Data Book, 22nd Edition*, September 2002; US Census Bureau, *The Journey to Work 2000*, March 2004.

18. UK Department for Transport, *National Travel Survey: 2004*.

19. Sustain, 'Eating Oil: Food Supply in a Changing Climate', www.sustainweb.org.

20. Department for Environment, Food and Rural Affairs, *Agriculture in the United Kingdom 2004*.

21. Professor David Pimentel, Cornell University, personal communication.

22. DEFRA, *Agriculture in the United Kingdom 2004*. These percentages are calculated on financial values rather than physical volumes.

23. Donald Winch, *Malthus*, OUP, 1987.

24. Garrett Hardin, *Living Within Limits: Ecology, Economics and Population Taboos*, OUP, 1993.

25. A.E. Johnstone, *The Efficient Use of Plant Nutrients in Agriculture*, IACR Rothamsted and International Fertilizer Industry Association.

26. Professor Vaclav Smil, 'Long-Range Perspectives on Inorganic Fertilizers in Global Agriculture', International Fertilizer Development Centre, October 1999.

27. David Pimentel, personal communication; UN FAO, http://faostat. fao.org.

28. Tushaat Shah, personal communication.

29. David and Marcia Pimentel, *Food, Energy and Society*, Edward Arnold, 1979.

30. Rattan Lal *et al.*, *Food Security and Environmental Quality in the Developing World*, CRC Press, 2002.

31. David Pimentel, personal communication.

32. David Pimentel, personal communication.

Chapter 6. Long-term Liquidation

1. *Daily Telegraph*, 3 February 2006.
2. Financial Services Authority, *Final Notice to the 'Shell' Transport and Trading Company, p.l.c. ('STT')*, 24 August 2004.
3. Securities and Exchange Commission, *Securities and Exchange Act of 1934, Release No. 50233*, 24 August 2004, instituting cease and desist proceedings against Shell.
4. Financial Services Authority, *Final Notice to the 'Shell' Transport and Trading Company*.
5. Davis Polk & Wardwell, *Report to the Shell Group Audit Committee*, 31 March 2004.
6. Ibid.
7. Simmons & Company International, *Simmons Equity Monthly, Integrated Oils: Reserve Reporting Season*, December 2005.
8. Figures provided by PFC Energy, personal communication.
9. Analysts believe Chevron's oil output will rise in 2006 and beyond, following its acquisition of Unocal in late 2005, and other new projects it has in development.
10. Art Smith wrote a paper in 1991 called 'M. King Hubbert's Analysis Revisited: An Update of the Lower 48 Oil & Gas Resource Base', published in the 1992 *International Association for Energy Economics North American Conference Proceedings*, which showed that Hubbert's 1956 forecast was still remarkably accurate for the early 1990s. Hubbert's forecast implied that oil production in the lower forty-eight states of the US would be 1.9 billion barrels in 1991, whereas actual production turned out to be 2 billion. Not bad.
11. John S. Herold, *Global Upstream Performance Review 2005*.
12. 'Oil Spike Gives BP Room to Return up to $65 Billion', *Financial Times*, 7 February 2006.
13. Eddy Isaacs, *Canadian Oil Sands: Development and Future Outlook*, www.aeri.ab.ca; EUCAR, JRC, CONCAWE, *Well-to-wheels Analysis of Future Automotive Fuels and Powertrains in the European Context, Well-to-Tank Report, Version 1*, December 2003.
14. *Oil and Gas Journal*, 19 December 2005, www.ogjonline.com.
15. Brian Straub, interviewed by the author, November 2005.
16. It is the huge energy demand of Steam Assisted Gravity Drainage that largely explains the low energy return of oil sands production – the fact that producing a barrel of synthetic crude on average consumes the equivalent of 33 per cent of the energy in that barrel.

17. Alberta Chamber of Resources, 'Oil Sands Technology Roadmap: Unlocking the Potential', 30 January 2004, http://www.acr-alberta. com/.

18. Len Flint, interviewed by the author, August 2005.

19. Isaacs, *Canadian Oil Sands*; Alberta Chamber of Resources, 'Oil Sands Technology Roadmap'.

20. IHS Energy; IEA, *World Energy Outlook 2005: Middle East and North Africa Insights*; BP *Statistical Review of World Energy*, 2005.

21. EUCAR, JRC, CONCAWE, *Well-to-wheels Analysis of Future Automotive Fuels and Powertrains in the European Context, Version 2A*, December 2005, www.ies.jrc.cec.eu.int/wtw.html.

22. IEA, *World Energy Outlook 2005: Middle East and North Africa Insights*.

23. International Energy Agency, *World Energy Outlook, 2006*.

24. EUCAR, JRC, CONCAWE, *Well-to-wheels Analysis of Future Automotive Fuels and Powertrains in the European Context, Version 2A*.

25. Amos Salvador, *Energy: A Historical Perspective and 21st Century Forecast*, AAPG Studies in Geology #54, 2005.

26. Jean Laherrère, *World Natural Gas Forecasts*, 5 March 2006, personal communication.

27. The largest LNG tankers contain 210,000 cubic metres of LNG, which regassifies to six times that volume, and the average UK household consumes 2,000 cubic metres per year. 210,000 x 600 / 2000 = 63,000 homes. Numbers provided by National Grid, which owns the Isle of Grain LNG terminal in Kent.

28. 'Qatar Seeks New Math for North Field', *World Gas Intelligence*, 1 June 2005.

29. *Simmons Oil Monthly – Qatar*, 24 April 2006, http://www.simmonsco-intl.com/files/Qatar%20Report.pdf.

30. Deutsche Bank Securities, *Global LNG: Waiting for the Cavalry*, 5 December 2005.

31. IEA, *World Energy Outlook 2006*.

32. 'Putting Glasnost in the Pipeline: The Rationale Behind BP's Alliance with Sidanco', *Financial Times*, 21 November 1997.

33. 'BP: From Hate to Love in Russia', *Sunday Times*, 5 October 2003.

34. 'BP is Still Wary of Russian Motives', *Guardian*, 17 October 2003.

35. BP quarterly accounts, www.bp.com.

36. Robin West, interviewed by the author, February 2006.

37. Stephen O'Sullivan, interviewed by the author, March 2006.

38. *BP Statistical Review*, www.bp.com

39. 'Russia Aims to Produce 510M Tonnes of Oil Annually by 2010 – Energy Minister', *MosNews*, 25 October 2005.

40. 'Total Chief Calls for Greater Access to OPEC Reserves', *The Times*, 21 February 2005.
41. 'Total Woos Alberta, but Love is Mideast', *Globe and Mail*, 24 September 2005.

Chapter 7. The Riddle of the OPEC Sands

1. *BP Statistical Review*, www.bp.com.
2. The IEA Saudi production total of 17.6 million barrels a day in 2030 is made up of 14.6 mb/d of crude, 2.7 mb/d of natural gas liquids and 0.3 mb/d from the Neutral Zone, where Saudi shares the output with Kuwait. IEA, *World Energy Outlook 2006*.
3. *BP Statistical Review*, www.bp.com. OPEC's members are: Algeria, Indonesia, Iran, Iraq, Kuwait, Libya, Nigeria, Qatar, Saudi Arabia, the United Arab Emirates and Venezuela. Angola joined on 1 January 2007.
4. Matthew R. Simmons, *Twilight in the Desert: The Coming Saudi Oil Shock and the World Economy*, John Wiley & Sons, 2005.
5. Despite the fact that they are evidently wrong, for the sake of consistency I will stick with the OPEC numbers, as reproduced in the *BP Statistical Review*.
6. Dr Colin Campbell, interviewed by the author, February 2005.
7. 'Kuwait Data Raise Reserve Level Questions', *Petroleum Intelligence Weekly*, 23 January 2006, and 'Oil Reserves Accounting: The Case of Kuwait', *Petroleum Intelligence Weekly*, 30 January 2006.
8. BP annual results press conference, 7 February 2006.
9. Launch of *BP Statistical Review 2006*, 14 June 2006.
10. Personal communication, Ken Chew, IHS Energy.
11. Edward Price, interviewed by the author, September 2005.
12. Europe has produced 58 billion barrels, of which the North Sea (UK, Norway, Denmark, Germany, Netherlands) accounts for 44 billion barrels. Personal communication, Ken Chew, IHS Energy.
13. Matt Simmons, interviewed by the author, September 2005.
14. You can hear the presentations at http://www.saudi-us-relations.org/energy/saudi-energy-abdul-baqi.html.
15. The USGS study is the one referred to in chapters 1 and 3. The study covered the entire world over the period 1996 to 2025. It estimated a total of 649 billion barrels of producible reserves (equivalent to 2P, or ultimate) remained to be discovered. If they were right, and if the oil was to be discovered by 2025, we should have been discovering it at a rate of 22 billion barrels a year. In fact we have been discovering less than 9 billion per year on average. This suggests that the USGS estimate of remaining resources is

in fact far too high – certainly in any meaningful timeframe. See Klett, Gautier and Ahlbrandt, 'An Evaluation of the U.S. Geological Survey World Petroleum Assessment 2000'.

16. Presentation by Mike Rodgers of PFC Energy at the Oil and Money Conference in London, 26 October 2004.

17. IEA, *World Energy Outlook 2005: Middle East and North Africa Insights.*

18. 'Saudi Oil Expansion Plan May Face Delay – Expert', Reuters, 10 November 2005.

19. 'Saudis Insist Capacity Plans on Track', *International Oil Daily*, 15 November 2005.

20. 'Forecast of Rising Oil Demand Challenges Tired Saudi Fields', *New York Times*, 24 February 2004.

21. 'OPEC Can't Meet West's Oil Demand, Say Saudis', *Financial Times*, 7 July 2005.

22. Sandra Mackey, *The Saudis: Inside the Desert Kingdom*, Norton, 2002.

23. US Department of Energy, Saudi Arabia Country Analysis Brief, www.doe.iea.gov.

24. Mackey, *The Saudis.*

25. Cited in Michael Klare, *Blood and Oil: How America's Thirst for Petrol is Killing Us*, Hamish Hamilton, 2004.

26. Ibid.

27. Chris Skrebowski, interviewed by the author, January 2006.

28. 31 January 2006, http://www.whitehouse.gov/stateoftheunion/2006/index.html.

29. US Department of Energy, http://tonto.eia.doe.gov.

30. 'The State of Energy', *New York Times*, 1 February 2006; 'A Union Address That Ran On Empty', *Financial Times*, 2 February 2006.

31. Robin West, interviewed by the author, February 2006.

Chapter 8. Interesting Times

1. 'Attack on Saudi Oil Facility Thwarted', Associated Press, 25 February 2006; 'Al Qaeda Claims Responsibility for Saudi Attack', Reuters, 24 February 2006; 'Al Qaeda Vows More Attacks After Saudi Oil Raid', Reuters, 25 February 2006.

2. Seymour M. Hersh, 'The Iran Plans: Would President Bush go to War to Stop Tehran from Getting the Bomb?', *New Yorker*, 17 April 2006, posted at www.newyorker.com on 8 April 2006.

3. 'Iran's Wargames See Oil Futures Rise by $2', *Financial Times*, 4 April 2006.

4. 'Iran's Message to the West: Back Off or We Retaliate', *Guardian*, 2 February 2006.

5. Yergin, *The Prize*, Simon & Schuster, 1991.

6. US-China Economic and Security Review Commission, *Annual Report*, November 2005, www.uscc.gov.

7. Statement before the House Committee on Armed Services, 13 July 2005 http://www.uscc.gov/testimonies_speeches/testimonies/2005/ 05_07_13_testi_damato.pdf.

8. US-China Economic and Security Review Commission, *Annual Report*, November 2005, www.uscc.gov.

9. Michael Klare, 'Oil, Geopolitics, and the Coming War with Iran', 11 April 2005, http://www.commondreams.org/views05/0411-21.htm.

10. '300 Retirement Funds in Danger of Collapse, Warns Watchdog', *Guardian*, 28 April 2006.

11. Robert L. Hirsch, Roger Bezdek, and Robert Wendling, *Peaking of World Oil Production: Impacts, Mitigation, and Risk Management*, February 2005, www.netl.doe.gov/publications/others/pdf/Oil_ Peaking_NETL.pdf.

12. James Turk and John Rubino, *The Coming Collapse of the Dollar and How to Profit From It*, Doubleday, 2004.

13. 2005 figures from Chris Sanders, of Sanders Research Associates.

14. 'IMF Warns High Prices Risk Global Crisis', *Financial Times*, 7 April 2006.

15. To be precise, $987.9 billion. Sanders Research Associates.

16. Dan Adleman, 'The Rise of the Petro Euro', *Republic*, 16 March 2006, www.republic-news.org.

17. Chris Sanders, interviewed by the author, April 2006.

18. In early 2002 the dollar was worth 0.8 Special Drawing Rights (a synthetic currency based on the value of a basket of real ones and gold), and in early 2005 it was worth only 0.65 SDRs. The dollar regained about a third of that loss during 2005, but seemed to have turned south again in 2006.

19. Javad Yarjani, 'The Choice of Currency for the Denomination of the Oil Bill', Oviedo, Spain, 14 April 2004. This speech is no longer on the OPEC website. Mr Yarjani is now No. 3 in the Iranian delegation to OPEC.

20. 'Dollar Falls on Chinese Diversification Fears', *Financial Times*, 4 April 2006; 'Asian Banks Cut Exposure to Ailing Dollar', *Financial Times*, 7 March 2005; 'Dollar Falls After Sweden Echoes Global Switch to Euros', *Daily Telegraph*, 22 April 2006; 'Dollar Sinks as G7 Appeals for Renminbi Flexibility', *Financial Times*, 24 April 2006; Janet Bush, 'America's Foes Prepare for Monetary Jihad', *New Statesman*, 4 October 2004.

21. US consumption (2005) of 20.6 mb/d divided by population of 298 million gives daily per capita consumption of 0.069 barrels, multiplied by world

population of 6.464 billion makes 447 mb/d. Proved reserves of 1.2 trillion barrels divided by 447 million comes to 2,680 days, or 7.3 years. Sources: World Resources Institute, http://earthscan.wri.org; *BP Statistical Review*, www.bp.com.

22. 'GM and Ford Report Slide in December Sales', *Financial Times*, 4 January 2006; 'Western Car Jobs Hit the Road', *Financial Times*, 16 February 2006.

23. The whole episode is entertainingly related by the Canadian commentator Linda McQuaig in her book, *It's the Crude, Dude: War, Big Oil, and the Fight for the Planet*, Doubleday Canada, 2004.

24. Ibid.

25. Ibid.

26. Center for Responsive Politics, www.opensecrets.org.

27. 'GM and Ford Report Slide in December Sales', *Financial Times*, 4 January 2006.

28. 'Western Car Jobs Hit the Road', *Financial Times*, 16 February 2006.

29. 'North American Auto Division Lost $1.6 bn in 2005', *Washington Post*, 24 January 2006.

30. Interviewed on BBC radio, 28 December 2004.

31. 'Airlines to Climb out of Six-year Loss-making Run', *Financial Times*, 22 March 2006.

32. 'Flying on Empty', *Economist*, 21 May 2005.

33. 'No 1 Retailer Blames Higher Gas Prices for Soft Second-quarter Sales, Worries About Outlook', CNN, 16 August 2005, www.cnn.com.

34. www.wakeupwalmart.com/facts/.

35. Deutsche Bank Securities, *Global LNG: Waiting for the Cavalry*, 5 December 2005.

36. Ziff Energy Group, 'Canadian Gas Exports to 2014', 31 March 2006, www.ziffenergy.com.

37. 'EU Says Gazprom Comments Confirm Dependency Fears', Reuters, 20 April 2006, www.reuters.com.

38. 'Russia Signs Deal to Build Gas Pipelines to China', *China Daily*, 22 March 2006, http://www.chinadaily.com.cn/china/2006-03/22/content_548962.htm.

Chapter 9. Memo to Mr Wicks

1. 'Can Johnson Cut It?', *Observer*, 15 May 2005.

2. Malcolm Wicks, speaking at the annual conference of the Parliamentary Renewable and Sustainable Energy Group (PRASEG), 12 July 2005.

3. Output fell 30 per cent between 1999 and 2005, when calculated as an average daily rate over the course of a full year, as it is reported in the *BP Statistical Review*. If you take the *monthly* production figures reported by Royal Bank of Scotland *Oil and Gas Index*, the decline from April 1999 to July 2005 was in fact 38 per cent. http://www.rbs.com/media 03b.asp?id=MEDIA_CENTRE/RBS_AND_THE_ECONOMY.

4. Klett, Gautier and Ahlbrandt, 'An Evaluation of the U.S. Geological Survey World Petroleum Assessment 2000'.

5. www.eia.doe.gov/pub/oil_gas/petroleum/feature_articles/2004/worldoilsupply/oilsupply04.html.

6. IEA, *World Energy Outlook 2006*.

7. Dr Richard Miller, interviewed by the author, March 2005.

8. BGR, *Reserves, Resources and Availability of Energy Resources*, 2004.

9. DIREM, *L'Industrie Pétrolière en 2004*, www.industrie.gouv.fr/energie.

10. Chris Skrebowski, interviewed by the author, April 2006.

11. Deffeyes, *Beyond Oil*.

12. Sir David King, speaking at the summer lunch of the Energy Institute, London, 12 July 2005.

13. Sir David Manning, 'Energy: A Burning Issue for Foreign Policy', Stanford University, 17 March 2006, http://www.britainusa.com.

14. Donald F. Fournier and Eileen T. Westervelt, *Energy Trends and Their Implications for US Army Installations*, US Army Corps of Engineers, Engineer Research and Development Center, September 2005.

15. 'Seizing the Nuclear Nettle', *Observer*, 4 December 2005.

16. David King, 'The Nuclear Option isn't Political Expediency but Scientific Necessity', *Guardian*, 16 December 2005.

17. 'So Now Green is the New Blue – and Also the New Red', *Observer*, 23 April 2006.

18. Ibid.

19. Ibid.

20. Roger Bentley, interviewed by the author, January 2005.

21. 'Peak Oil Letter from UK Energy Minister', *Energy Bulletin*, 9 May 2006, http://www.energybulletin.net/15775.html.

Chapter 10. Passnotes for Policymakers

1. 'Total Sees 2020 Oil Output Peak, Urges Less Demand', Reuters, 7 June 2006.

2. 'PM's Change of Heart on Nuclear Power Issue', *Independent*, 5 July 2006.

3. 16 May 2006, http://www.pm.gov.uk/output/Page9470.asp.

4. House of Commons Environmental Audit Committee, *Keeping the Lights On: Nuclear, Renewables and Climate Change*, Sixth Report of Session 2005–06, Vol. 1, HC584–1, 16 April 2006,
   http://www.publications.parliament.uk/pa/cm/cmenvaud.htm

5. Ibid.

6. Platts, 'Nuclear Growth Faces Supply-side Constraints', 15 August 2006,
   www.platts.com/nuclear/resources/news%20features/nukegrowth/index.xml.

7. Dr Sam Holloway, personal communication, 8 June 2006. Total UK emissions in 2004 were $c.575$ million tonnes CO_2, of which the twenty largest emitters put out $c.132$mt. Some adjustment needs to be made for the fact that running CCS consumes additional energy, and the process only captures about 85 per cent of the CO_2 emitted. This means that in order to sequester 132mt of generating emissions, you would actually need to bury 140mt. Holloway estimates (roughly) that British North Sea depleted gas fields could hold 6,600mt, and the oilfields 1,170mt. Total capacity of 7,770mt divided by 140mt/year gives $55\frac{1}{2}$ years. The saline aquifer capacity of the southern North Sea is almost double, at 14,000mt, but Holloway stresses that although the structures have been identified, since they contain only water and no buoyant fluids, it is impossible to tell without further investigation whether or not they would leak.

8. According to the *BP Statistical Review 2005*, the R/P ratio for world proved coal reserves is 164 years. Of course the rate of coal consumption is soaring, so these reserves are likely to be exhausted far sooner. In other words, we seem to have plenty of CCS capacity in the UK to be getting on with.

9. House of Commons Environmental Audit Committee, *Keeping the Lights On*.

10. D. J. Milborrow, D. J. Moore, N. B. K. Richardson and S. C. Roberts, *The UK Offshore Windpower Resource*, Central Electricity Generating Board, 1982.

11. H. G. Matthies *et al.*, *Study of Offshore Wind in the EC*, 1995. JOUR 0072, Vertga Naturliche Energie and repeated in ETSU report W/35/00250/REP/1, http://www.bwea.com/business/market.html.

12. Report by PB Power for the Royal Academy of Engineering, *The Costs of Generating Electricity*, March 2004, available at http://www.raeng.org.uk/news/publications/list/default.htm?TypeID=2.

13. Wind produced 1,935GWh in 2004, while nuclear produced 71,820GWh, 37 times as much. Source: House of Commons Environmental Audit Committee, *Keeping the Lights On*. To calculate how many wind turbines would be needed to replace the 71,820GWh electricity generated by nuclear in 2004: 72,000GWh / 8,760 (hours in the year) gives nominal

generating capacity required of 8.2GW. But onshore wind turbines have an average load factor of around 30 per cent, so this capacity must be multiplied by 3.33, to give 27GW wind capacity required, or 27,000MW. So assuming the turbines were 3MW each, you would need 9,000 of them. Britain had installed 1,682 turbines by August 2006; source: www.bwea.com/ukwed/index.asp.

14. House of Lords Science and Technology Committee, *Energy Efficiency*, 15 July 2005, http://www.parliament.the-stationery-office.co.uk/pa/ld/ldsctech.htm.

15. Ibid.

16. George Monbiot, 'Strange but True – Shoddy Building Work in Exeter Kills People in Ethiopia', *Guardian*, 30 May 2006, www.guardian.co.uk/climatechange/story/0,,1785714,00.html.

17. House of Lords Science and Technology Committee, *Keeping the Lights On*.

18. Greenpeace, *Decentralising Power: An Energy Revolution for the 21st Century*, launched at London City Hall, 19 July 2005.

19. Woking Borough Council, 'Energy, Climate Neutral Development, A Good Practice Guide', http://www.woking.gov.uk/council/planning/publications/climateneutral2/energy.pdf.

20. UK gas consumption was 85.1mt of oil equivalent in 2005 (*BP Statistical Review*), or 3.9bn GJ (1 tonne of oil equivalent = 45.447980266 GJ, according to *Bossley's Energy Conversions*). Let's assume that Britain goes headlong for CHP, and that (however unlikely) this cuts gas consumption by 30 per cent and keeps it there, meaning consumption is reduced to 2.7bn GJ. Britain has 5.66 million hectares of cropland (www.fao.org), and let's say all of this was turned over to short rotation coppice (willow), to produce wood for gasification. This crop would yield between 160 and 250 GJ per hectare (IEA, personal communication), producing a maximum of 1.415bn GJ if planted on all cropland. The gasification process is 85 per cent efficient, meaning that the net energy production (gas available for CHP and other use) would in fact be 1.202bn GJ, less than half that required, even with consumption reduced by 30 per cent – and we'd starve. Willow and other such crops can be grown on poorer quality land, and Britain has twice as much pasture as cropland (11,248,000 hectares), but even if all the pasture were put to willow, it would still fail to produce enough energy: 11,248,000 × 250 / 0.85 = 2.39bn GJ. And there would be no free-range meat or milk, and nowhere to grow any biofuels for road transport. According to Dr Jeremy Woods of Imperial College, who is enthusiastic about the potential for biofuels globally, the same is probably true of any densely populated, temperate country.

21. Permit Glut Undermines EU Carbon Scheme, *Financial Times*, 16 May 2006.
22. Colin Challen, interviewed by the author, May 2006.
23. Matt Prescott, interviewed by the author, June 2006.
24. BBC Radio 4 News, 6 June 2006.
25. 'US Cities Snub Bush and Sign up to Kyoto', *Guardian*, 17 May 2005; the carbon trading scheme is organized by the US Regional Greenhouse Gas Initiative, www.rggi.org.
26. See the full protocol at the ASPO Ireland website, http://www. peakoil.ie/protocol.
27. Colin Campbell, interviewed by the author, August 2004.
28. Mayer Hillman, *How We Can Save the Planet*, Penguin, 2004.
29. Ibid.
30. Gordon Brown speaking at the launch of the Stern Review, 30 October 2006, http://www.hm-treasury.gov.uk/newsroom_and_speeches/speeches/chancellorexchequer/speech_chx_301006.cfm; Tony Blair cited in 'Climate Action Consensus Elusive', *Financial Times*, 31 October 2006.
31. Stern Review on the economics of climate change, http://www.hm-treasury.gov.uk/Independent_Reviews/stern_review_economics_climate_change/sternreview_index.cfm.

Chapter 11. How to Survive the Imminent Extinction of Petroleum Man

1. 'How to be a Greener Driver', *Independent*, 21 November 2005; also http://www.newscientist.com/article/mg18925432.200.html.
2. http://www.fueleconomy.gov/feg/noframes/19813.shtml; http://www.smart.com; 'How Green Can You Get?', *Independent*, 7 February 2006.
3. 'Natural Selections', *Independent*, 20 June 2006.
4. Dr Ben Lane, an environmental consultant, has calculated the per-mile fuel costs of a range of different cars and fuel types, valid for mid-2006: 1p for the G-Wiz; 7.5p for the Toyota Prius petrol hybrid; and 10p for a petrol car in vehicle excise band D. www.ecolane.co.uk.
5. Dr Ben Lane, interviewed by the author, June 2006.
6. *UK National Travel Survey 2004*, table 5.2, http://www.dft.gov.uk/stellent/groups/dft_transstats/documents/pagc/dft_transstats_039332.xls.
7. 'True Blue Green', *Guardian*, 3 December 2005.
8. Hillman, *How We Can Save the Planet*.

9. 'Lower Bills May Not be Blowing in the Wind', *Observer*, 25 June 2006.

10. *Bossley's Energy Conversions*, Petroleum Economist, 2002.

11. 9.6W × 8,760 (hours in the year) = 84 kWh. The G-Wiz charges in 8 hours from a 2.2kW charger, meaning the car consumes 17.6kWh to travel its full range of 48 miles. 17.6 goes into 84 almost 4.8 times. 48 miles × 4.8 = 229.

12. 31W (TV cluster) + 53W (computer cluster) = 84W. 84W × 8,760 (hours in year) = 735kWh. 735 / 17.6kWh (per G-Wiz recharge) = 41.8. 41.8 recharges × 48 miles (G-Wiz range) = 2,006 miles. All right, all right, according to the *Michelin Motoring Atlas Europe*, London to Athens is actually 3,252km/2,020 miles, so you'd have to walk the last bit. But if I threw a few more of my appliances into the equation you would not only get to Athens, but over a third of the way home again. Including the Roberts radio (9.6W), the cordless telephone (9.7W), and 14.5W drawn by a couple of sneaky gadgets hidden in my hi-fi system that have no off buttons, the total potential power wasted is almost 118W − or the equivalent of 2,814 miles.

13. Norman Baker MP, *How Green is Your Supermarket?*

14. International Council of Bottled Water Associations, www.icbwa.org/ 2000-2003_Zenith_and_Beverage_Marketing_Stats.pdf.

15. Presentation by BP chief economist Peter Davies, launching *BP Statistical Review 2006*, 14 June 2006.

16. Lester Brown, 'World Grain Stocks Fall to Record Low', 20 June 2006, http://www.peopleandplanet.net/doc.php?id=2784; 'Ethanol's Demand for Corn May Trim US Meat Output', Reuters, 23 May 2006.

17. 'Revolution! Britain Embraces the Bicycle', *Independent*, 7 June 2006.

Internet Resources

Links to all the following websites and more can be found at www.lastoilshock.com

Oil industry and environmental news sites

http://www.planetark.com/
http://www.platts.com/
http://realtimenews.slb.com/news/
http://www.gasandoil.com/goc/welcome.html

Peak oil organizations

http://www.odac-info.org/
The UK Oil Depletion Advisory Centre, which sends out a useful newsletter by email.

http://www.peakoil.net/
The Association for the Study of Peak Oil, founded by Colin Campbell, and now with branches in many countries around the world, also with a useful newsletter.

http://www.powerswitch.org.uk/
Powerswitch is a British organization dedicated to raising public awareness of peak oil and promoting self-help.

http://www.hubbertpeak.com/
Excellent site for those interested in reading Hubbert first hand. There
is a good bibliography, and links to many of the important papers and
academic articles about his work.

http://www.energybulletin.net/
http://www.drydipstick.com/
http://www.globalpublicmedia.com/

Acknowledgements

AFTER EIGHTEEN MONTHS of research and over 170 interviews it would be impossible to thank individually everybody who helped me to write this book. I am deeply indebted to scores of energy industry professionals and academics who were generous with their time and insights, including many who are not quoted directly in the work but whose contributions were nevertheless vital in helping me understand the subject.

In particular I thank Roger Bentley, Chris Skrebowski, Trevor Ridley, Dr Jeremy Woods, Jason Nunn, John Olsen, Chris Sanders, Professor Joy Gordon, Kevin McConway, James Blewett, Rob Hayward, and one source deep in the heart of the international oil industry who will remain nameless to protect the innocent. They all spent valuable time either patiently explaining key concepts in words of one syllable, discussing technical and political issues, or reviewing parts of the manuscript. A number of other respondents also offered invaluable feedback, but of course any remaining howlers are entirely my fault.

Ken Chew was generous not only with his time and understanding of some of the most tortuous questions of oil industry statistics, but also provided lots of useful numbers from the IHS Energy database. I am also grateful for data from Argus (www.argusmediagroup.com), PFC Energy, John S. Herold, Simmons & Company International, Dr Michael Smith of www.energyfiles.com, BOC, and the IEA. Oil and gas markets expert Liz Bossley of CEAG kindly provided the indispensable *Bossley's Energy Conversions* (Petroleum Economist, 2002).

Fellow journalists who were generous with archive articles and source material include Rob Evans of the *Guardian*, Peter Kemp and

ACKNOWLEDGEMENTS

Barbara Shook of *Energy Intelligence*, Barbara Lewis of Reuters, David Knott and Bill Farren-Price of the *Middle East Economic Survey*, and Janet Bush. Others who did favours, opened doors or generally made things happen include Paul Tempest, Christopher Farrell, Ron Swenson, my sister Harriet Griffey, and Sue Webb and Debbie Enright of Network Typing Services. Bill Garrett saved my computer – and bacon – on several occasions, and Ciara Cannon did heroic work on the graphics.

This book might never have been written had it not been for Jonny Geller who jumped on the idea, and so became my agent, and editors Gordon Wise, Eleanor Birne and Helen Hawksfield who commissioned it and gently steered me through the daunting task of writing my first book. My thanks to them all.

Index